INTERNATIONAL CENTRE FOR MECHANICAL SCIENCES

COURSES AND LECTURES - No. 41

WLODZIMIERZ PROSNAK

COMPUTING CENTER POLISH ACADEMY OF SCIENCES, WARSAW

METHOD OF INTEGRAL RELATIONS

THEORY AND SELECTED APPLICATIONS
TO BLUNT - BODY PROBLEMS

COURSE HELD AT THE DEPARTMENT
OF FLUIDDYNAMICS
JULY 1970

UDINE 1970

SPRINGER-VERLAG WIEN GMBH

Originally published by Springer-Verlag Wien New York in 1972

ISBN 978-3-211-81146-7   ISBN 978-3-7091-2858-9 (eBook)

DOI 10.1007/978-3-7091-2858-9

INTRODUCTION

I wish to begin with words of gratitude
to Professor L.SOBRERO, General Secretary of the CISM
and Professor L. NAPOLITANO, Director of the Department of Fluid Mechanics of the CISM for inviting me
and letting me speak on the subject which was for
several years the main field of interest for myself
and some of my collaborators.

This subject, the method of integral
relations (MOIR), falls within the scope of the second
line of the present Course on Computational Gasdynamics, as it was pointed out by Professor NAPOLITANO in
his first lecture, the "Fundamentals" and "Applications" being the first and third line, resp. It belongs to general "instruments" or techniques - if you
prefer - that are available for solving problems
formulated for the partial differential equations of
classical gasdynamics. However, it should be empha-
sized perhaps already at this point, that the
method - although developed for specific needs and
problems of fluid mechanics is no less general an
instrument, than the method of finite differences -
for example. It can be applied to any problems posed
for systems of partial differential equations, provided
that the equations themselves could be reduced to the
so-called "divergence" form.

There is not much use in presenting

_solely a method of solution, and giving no examples
of its application, as only applications allow one to
form an opinion concerning efficiency of the method.
Therefore I shall include_ some examples _into my lectures : those concerning the blunt body problems,
because - as I understand - this is what the audience
is mainly interested in._

The material of the lectures _will be
divided into three chapters, the first one being devoted to general theory of the MOIR, the second one -
to application of the MOIR to typical blunt body problems, and the third one - to some other selected problems of blunt body flows._

The general purpose _of the lectures is
to prepare the audience to unaided solving of gasdynamic problems by means of the MOIR, and to demonstrate possibilities as well as limitations of this
method by discussing in detail some problems already
solved. Therefore I will confine myself - especially
in the first chapter - only to the most simple problems, so that the main ideas would not be overshadowed
by large amount of details of "second order of impor-
tance". However, keeping in mind my intention of
teaching, how to use the MOIR in practice, I probably
shall not be able to avoid some boring details entire-
ly._

The fundamental concept _of the MOIR
consists in transformation of a system of partial differential equations into a system of ordinary diffe-_

rential equations. Of course, the boundary conditions
for the first of these two systems have to be transformed into suitable conditions for the second one,
too.

However, solution of the resulting
system of ordinary differential equations obtained by
means of the MOIR, does not belong to the MOIR itself.
It depends on the system, and can be achieved by
analytical, approximate or numerical methods. Usually,
the resulting system is nonlinear and rather complicated, so that it has to be solved by the use of
numerical methods, and electronic digital computers.

The main advantage in using the MOIR
is equivalent - both from the numerical as well as
from the theoretical viewpoint - to the difference in
dealing with problems for systems of ordinary differen
tial equations instead of problems for systems of
partial differential equations.

This difference becomes especially
pronounced when the lastmentioned systems are nonlinear, of high order and of high number of independent
variables, which is precisely the case in fluid mechanics, and even in classical gas dynamics, i.e. dynamics of nonviscous, not heat-conducting gas, obeying
the Clapeyron's law. These properties of the system
of equations of the fluid mechanics explain perhaps
the fact, why the MOIR has been first developed for
specific needs and purposes of fluid mechanics, and
why - despite its universlity - the various branches

of this science still remain its most common field of
application.

It is perhaps advisable not to prolong
these general remarks and rather to save them until
the end of the first chapter, when the audience will
be already more familiar with the MOIR itself. Therefore I will close this introduction adding only a few
words concerning history of the MOIR.

All the fundamental ideas of the MOIR
can be found in the Pohlhausen's paper [1] published
in 1921, and the method itself has been known since
that date, however, only in its application to boundary layer problems. In the late fifties Dorodnicyn
[2,3] presented the MOIR in quite new and very general
formulation. To him belongs also the whole credit of
starting new applications of the method, especially
to mixed problems of classical gas dynamics. These
and similar applications have been cultivated mainly
in U.S.S.R. and U.S.A. during the last decade, and -
as I said at the beginning - some work on the MOIR
was done also in Poland. We were concerned with limitations of the method, and experimenting with its applications to nontypical problems rather then with
production of large number of numerical solutions for
different body shapes, different Mach numbers and so
on. Some of our results will be presented in this
course.

Udine, July 1970

Chapter 1
GENERAL THEORY

1.1 Derivation of the basic formulae.

We shall attempt now to present the basic pattern
of the MOIR, using for the sake of simplicity a system of partial
differential equations with only two independent variables x and
y. In presenting this pattern we will follow to some extent
the Dorodnicyn's approach [3].

Let's assume that we deal with a problem, posed
for a system of partial differential equations, and suitable
boundary conditions. The system is "closed" and it contains n
unknown functions $u_1, u_2, \ldots u_n$ of the independent variables x and
y. Let's assume, that the system can be presented in the
"divergence" form :

$$\frac{\partial P_i}{\partial x} + \frac{\partial Q_i}{\partial y} = F_i \; ; \; i = 1, 2, 3 \ldots n, \qquad (1.1)$$

which is always the case as far as equations of classical gasdynamics are concerned.

The meaning of the symbols in (1.1.) is as follows :

$$P_i = P_i(x, y; \quad u_1, u_2, \ldots, u_n);$$

$$Q_i = Q_i (x, y ; u_1, u_2, \ldots, u_n) ;$$

$$F_i = F_i (x, y ; u_1, u_2, \ldots, u_n) ;$$

P_i, Q_i, F_i being known functions of their arguments. It is obvious, that the system (1.1) consists of first order differential equations only : after differentiation of the functions P_i, Q_i only the first derivatives of u_i would appear in every equation.

The solution to the system (1.1) be sought in rectangular region :

(1.2)
$$\begin{cases} a \leq x \leq b ; \\ c \leq y \leq d . \end{cases}$$

The conditions, which have to be satisfied by the functions u_i on the sides of this rectangle, be as follows :

(1.3)
$$\begin{cases} G_\nu (a, y ; u_1(a, y), u_2(a, y), \ldots, u_n(a, y)) = 0 ; \ \nu = 1, 2 \ldots R \\[2mm] G_\nu (b, y ; u_1(b, y), u_2(b, y), \ldots u_n(b, y)) = 0 ; \ \nu = R+1, R+2, \ldots n ; \\[2mm] G_\nu (x, c ; u_1(x, c), u_2(x, c), \ldots, u_n(x, c)) = 0 ; \ \nu = 1, 2, \ldots S, \\[2mm] G_\nu (x, d ; u_1(x, d), u_2(x, d), \ldots u_n(x, d)) = 0 ; \ \nu = S+1, S+2, \ldots n \end{cases}$$

We assume the conditions being such that the problem for the system (1.1) is properly posed from the mathematical viewpoint.

G_υ denotes known functions of their arguments. In special cases the rectangle (1.2) may degenerate to semiinfinite band :

$$b = +\infty \qquad \text{or} \qquad c = +\infty \qquad (1.4)$$

as well as to infinite one :

$$a = -\infty\,, \qquad b = +\infty\,; \quad \text{or} \quad c = -\infty\,, \qquad d = +\infty\,. \quad (1.5)$$

The boundary conditions (1.3) must assume in such cases a corresponding form.

The fundamental concept of the MOIR - that of transforming a system of partial differential equations into an approximate system of ordinary differential equations, is realized first of all by performing the following two closely interconnected operations :

1st integration of the system (of partial differential equations) with respect to this particular independent variable, which has to vanish from the system ;

2nd suitable approximation of the integrals appearing after performing of the abovementioned integration.

In order to illustrate in the most simple manner, how it is done, let's assume, that y is the independent variable, which has to disappear. In accordance with this assumption let's divide the rectangle (1.2) into N "stripes" by equidistant straight lines :

(1.6) $$y = y_k = \text{const}$$

as it is shown in the Fig. 1.1.

Fig. 1.1

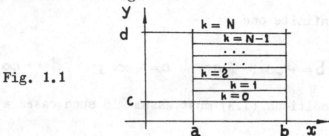

Let's introduce the notation :

(1.7) $$\begin{cases} F_{i,k} = F_i(x, y_k) \, ; & i = 1, 2, \ldots, n \, ; \\ Q_{i,k} = Q_i(x, y_k) \, ; & k = 0, 1, \ldots N, \\ u_{i,k} = u_i(x, y_k) \, , \end{cases}$$

the symbols $F_{i,k}$, $Q_{i,k}$, $u_{i,k}$ representing functions of one argument only, i.e. of x .

Now we perform the first of the two operations, i.e. we integrate the system (1.1) with respect to y across every strip, obtaining in such a way the following <u>system of integral relations</u> :

(1.8) $$\frac{d}{dx} \int_{y_k}^{y_{k+1}} P_i \, dy + Q_{i,k+1} - Q_{i,k} = \int_{y_k}^{y_{k+1}} F_i \, dy \, ,$$

where we substituted :

$$\int_{y_k}^{y_{k+1}} \frac{\partial P_i}{\partial x} \, dy = \frac{d}{dx} \int_{y_k}^{y_{k+1}} P_i \, dy \,, \tag{1.9}$$

as the limits of integration do not depend on x , in the case considered.

The second operation consits in expressing the integrals by means of the functions (1.7) of the one variable.

Using any integration formula known from the numerical analysis, we obtain :

$$\left. \begin{aligned} \int_{y_k}^{y_{k+1}} P_i \, dy &= (d-c) \sum_{\gamma=0}^{\gamma=M} A_\gamma P_i \,; \\ \int_{y_k}^{y_{k+1}} F_i \, dy &= (d-c) \sum_{\gamma=0}^{\gamma=M} A_\gamma F_i \,, \end{aligned} \right\} \tag{1.10}$$

the constants A_γ and M depending on the formula applied.

Using e.g. the trapezoidal rule one obtains :

$$\left. \begin{aligned} \int_{y_k}^{y_{k+1}} P_i \, dy &= \tfrac{1}{2} \left(P_{i,k} + P_{i,k+1} \right) \left(y_{k+1} - y_k \right) \,; \\ \int_{y_k}^{y_{k+1}} F_i \, dy &= \tfrac{1}{2} \left(F_{i,k} + F_{i,k+1} \right) \left(y_{k+1} - y_k \right) \,; \end{aligned} \right\} \tag{1.11}$$

where

$$y_{k+1} - y_k = \frac{1}{N} (d-c)$$

accordingly to the former assumption, concerning the equal
width of the strips.

It can be easily seen, that the operation discussed represents in fact the approximation of the integral of a
function of one variable y by means of discrete values of this
function at the points y_k , the expressions (1.7) playing the
role of the discrete values, and x -that of a parameter.

Substituting (1.10) or (1.11) into the system
of integral relation (1.8) and performing the differentiation
with respect to x , we arrive at <u>the system of ordinary differential equations</u>, containing $u_{i,k}(x)$ as unknown functions.

There are $n\,(N+1)$ unknown functions corresponding
to the $(N+1)$ dividing lines, but only $n \cdot N$ equations, corresponding to N strips. In order to "close" the system , the lacking
n equations are to be derived from the boundary conditions
(1.3), or - more precisely - from the third and forth expression
in the set (1.3).

<u>The boundary conditions</u> for the final system of
ordinary differential equations are formulated in obvious manner
by means of the remaining two expressions (1.3). In "lucky"
cases the boundary conditions reduce themselves to initial conditions, so that one arrives at a Cauchy's problem for the system of first order differential equations, which is especially
convenient in view of ready, standard numerical methods for
solving problems of this type.

Substituting e.g. (1.11) into (1.8) and taking into account (1.3) one obtains finally the following "closed" system of ordinary differential equations :

$$\frac{d}{dx}\left(P_{i,k}+P_{i,k+1}\right)+2\,\frac{Q_{i,k+1}-Q_{i,k}}{y_{k+1}-y_k}=\left(F_{i,k}+F_{i,k+1}\right)\;;$$

$$G_\nu\left(x,c\,;\,u_{1,0}\,,u_{2,0},\ldots u_{n,0}\right)=0,\;\nu=1,2\ldots,S\,;$$

$$G_\nu\left(x,d\,;\,u_{1,N}\,,u_{2,N}\ldots u_{n,N}\right)=0,\;\nu=(S+1),(S+2)\ldots n$$

(1.12)

with the following boundary conditions :

$$G_\nu\left(a,y_k\,;\,u_{1,k},u_{2,k}\ldots u_{n,k}\right)=0,\qquad\nu=1,2,\ldots R\,;$$

$$G_\nu\left(b,y_k\,;\,u_{1,k}\,,u_{2,k}\ldots u_{n,k}\right)=0,\;\nu=(R+1),(R+2)\ldots n$$

(1.13)

There are in total $n(N+1)$ conditions, which equals the number of unknown functions as well as the number of equations in the system (1.12).

The derivatives of the unknown functions $u_{i,k}$ may appear explicitely in (1.12) only after determination of the functions $P_{i,k}$ and performing the differentiations indicated.

The number of strips N is usually identified with the "order of approximation" of the original system (1.1).

1.2. Curved boundaries.

The lines dividing the strips have not necessarily to be straight, as it was shown in the Fig. 1.1. : in general case they may be curved (Fig. 1.2).

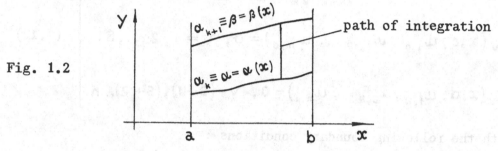

Fig. 1.2

Dealing with curvilinear boundaries requires the use of the Leibnitz's formula :

$$(1.14)\frac{d}{dx}\int_{\alpha(x)}^{\beta(x)} f(x,y)\,dy = \int_{\alpha(x)}^{\beta(x)} \frac{\partial f}{\partial x}\,dy + \frac{d\beta}{dx} f(x,\beta(x)) - \frac{d\alpha}{dx} f(x,\alpha(x)).$$

Comparing it with (1.9) we can see at once that two additional terms will appear now in every integral relation.

The curved boundary lines of the strips may be prescribed explicitely : in this case the functions $\alpha(x)$, $\beta(x)$ are given. Sometimes, as it is the case precisely in blunt body problems, some boundaries are not known in advance ; they constitute part of the whole solution to the problem. In such cases corresponding differential equations for the curved boundaries must be included into the system in order to close it.

1.3. Complementary conditions.

In the final system of ordinary differential equations, as obtained by applying of the MOIR, sometimes singularities occur. The type of such singularities (saddle point, nodal point and so on) has to be investigated thoroughly, and a condition concerning the behaviour of the solution in the vicinity of the singular point has to be imposed. It may stem from mathematical as well as from physical considerations.

Such a condition as well as any condition, which does not appear in the boundary conditions (1.3) of the original problem for the system of partial differential equations, but must be imposed on the solution to the final system of ordinary differential equations, will be called a complementary condition.

The regularity requirement in the Pohlausen's method can serve as an example of a "complementary" condition, as it will be shown in one of the examples. (see 1.6.2).

1.4. The consecutive operations.

The consecutive operations connected with applying the MOIR to a problem formulated for a system of partial differential equations, and described in the previous paragraphs, can be summarized as follows :

I. Transformation of the system of partial differential equations to the "divergence" form (1.1).

II. Choice of the most suitable form of approximation : selection of the independent variable, which has to vanish ; decision concerning number and shape of the strips. Short : selection of "the scheme of the approximation".

III. Integration with respect to the chosen independent variable, "across" every strip.

IV. Evaluation of integrals in accordance with the chosen "scheme of approximation", and substitution of the expressions obtained into the system of integral relations (1.8). (The curved boundaries, if such appear in the accepted scheme of approximation, must be properly taken into account).

V. Performing of the differentiations, so that the derivatives of the unknown functions appear explicitely.

VI. Providing additional equations in order to "close" the system, if necessary.

VII. Formulation of the appropriate conditions for the "closed" system of ordinary differential equations : initial, boundary, and complementary conditions.

In order to demonstrate, how these operations are performed in practice, a number of very simple examples will be presented.

In course of solving every example the operations as summarized in this paragraph will be reffered to by their Roman Numbers.

1.5. Example of a linear parabolic equation .

1.5.1. <u>Purpose</u>. – to show how the method works in a possibly simple case.

1.5.2. <u>Problem</u>. (conductivity) :

$$u_t = a^2 u_{xx} + f(t) ;$$

$$a = const,$$

$f(t)$ – given function,

$$u = u(x,t),$$

$$0 \leq x \leq l ; \qquad\qquad 0 \leq t < +\infty ;$$

initial condition : $u(x,0) = \varphi(x);$

boundary conditions : $u(0,t) = F_1(x);$

$$u(l,t) = F_2(t);$$

φ, F_1, F_2 –given functions.

1.5.3. <u>Performing of the consecutive operations</u>.

ad I – transformation to the "divergence" form :

$$u_t = a^2 \psi_x + f(x,t);$$

$$u_x = \psi$$

ad II – choice of the scheme of approximation :

N "strips" are chosen parallell to the t-axis, with border lines :

$$x_k = k \cdot \frac{l}{N} ; \qquad k = 0, 1, \ldots N ;$$

linear approximation across every strip is
assumed (trapezoidal rule).

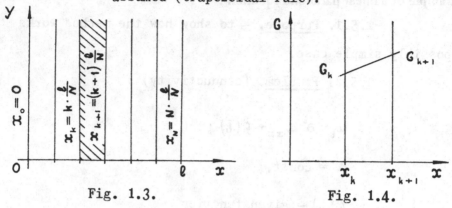

Fig. 1.3.   Fig. 1.4.

ad III - integration with respect to x :

$$(1.15)\begin{cases} \dfrac{\partial}{\partial t}\int_{x_k}^{x_{k+1}} u\,dx = a^2\,(\psi_{k+1}-\psi_k)+\int_{x_k}^{x_{k+1}} f(x,t)\,dx \\[4mm] u_{k+1}-u_k = \displaystyle\int_{x_k}^{x_{k+1}} \psi\,dx\,. \end{cases}$$

Notation : $u_k = u(x_k,t)$; $\psi_k = \psi(x_k,t)$.

ad IV - evaluation of integrals (Fig.1.4) :

$$(1.16)\qquad \int_{x_k}^{x_{k+1}} G\,dx = \tfrac{1}{2}(G_k+G_{k+1})(x_{k+1}-x_k)$$

- in accordance with the accepted trapezoidal
rule.

ad V - performing of the differentiations. After
differentiation of the expressions (1.16) with respect to t

in case when

$$6 \equiv u$$

and after substituting the results into (1.15),
the following system of ordinary differential
equations can be obtained :

$$\left. \begin{array}{l} \dfrac{du_k}{dt} + \dfrac{du_{k+1}}{dt} = 2a^2\left(\psi_{k+1} - \psi_k\right)/\left(x_{k+1} - x_k\right) + f_{k+1} + f_k \ ; \\[2mm] u_{k+1} - u_k = \dfrac{1}{2}\left(\psi_{k+1} + \psi_k\right)\left(x_{k+1} - x_k\right) , \end{array} \right\} \qquad (1.17)$$

consisting of N differential and N algebraic
equations with $2\left(N+1\right)$ unknown functions :

$$u_o, u_1, \dots u_N ; \qquad \psi_o, \psi_1, \dots \psi_N . \qquad (1.18)$$

ad VI – derivation of the two lacking equations
in order to "close" the system.
They follow from the boundary conditions of the
original problem, as presented 1.5.2. :

$$u_o \equiv F_1\left(t\right) ; \qquad u_N \equiv F_2\left(t\right) \qquad (1.19)$$

It can be easily seen, that ψ_k can be eliminated
from the system (1.17), and finally one obtains
the Cauchy's normal form :

$$\dfrac{du_k}{dt} = \phi_k\left(u_k ; F_1 , F_2\right); \quad k = 1, 2, \dots \left(N-1\right) \qquad (1.20)$$

the constants being omitted in (1.20).

ad VII – formulation of suitable conditions.

They are i n i t i a l conditions in the case
considered, following from the initial condition
of the original problem:

(1.21)
$$u_k(0) = \varphi(x_k) ; \qquad k = 1, 2, \ldots (N-1).$$

1.5.4. <u>Result</u> : by means of the MOIR the original
problem stated in 1.5.2. has been transformed into the Cauchy's
problem for the system (1.20) of ordinary differential equations
with initial conditions (1.21).

1.5.5. <u>Exercise</u> : derive the system (1.20) for
the special case of three "strips" : **N = 3** .

1.5.6. <u>Exercise</u> : derive a system similar to
(1.20) for the special case of two "strips", assuming, however,
the following boundary conditions:

$$u_x(0, t) = H_1(t) ;$$
$$u_x(\ell, t) = H_2(t)$$

in the original problem 1.5.2.

1.5.7. <u>Exercise</u> : apply the MOIR to the problem
described by the wave equation :

$$u_{tt} = a^2 u_{xx},$$
$$a = \text{const} ;$$
$$u = u(x, t),$$

$$0 \leq x \leq \ell \qquad ; \qquad 0 \leq t < +\infty,$$

and the following conditions :

$$u(x,0) = H_1(x); \qquad u_t(x,0) = H_2(x);$$

$$u(0,t) = u(\ell,t) = 0$$

1.5.8. <u>Exercise</u> : apply the MOIR to the problem described by the Poisson's equation :

$$\Delta u = -2$$

where

$$u = u(x, y);$$

Δ – Laplace operator,

in a simply connected region bounded by a line L, with the boundary condition

$$u = 0 \qquad\qquad\qquad\qquad (\text{on } L).$$

<u>Note</u> that there are at least three possibilities of presenting the equation in the "divergence" form, as consisting of an equivalent system of two first order partial differential equations.

1.6. Example of the boundary layer equation .

1.6.1. Purpose.

        - to show application of the MOIR in case of curvilinear bounds of the strips ;

        - to show approximation by a polynomial, instead of that by rectilinear segments, as in the previous example ;

        - to show how a complementary condition is derived from the system of ordinary differential equations ;

        - incidentally : to show that the Pohlhausen's method [1] represents essentially nothing else but a special case of the MOIR , as it includes all the operations summarized in the par. 1.4.

1.6.2. Problem (boundary layer) :

$$(1.22) \qquad u\, u_x + v\, u_y = U U' + v\, u_{yy} \; ;$$

$$(1.23) \qquad u_x + v_y = 0$$

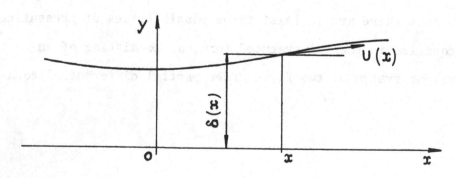

Fig. 1.5.

where :

$$u = u(x, y); \qquad v = v(x, y)$$

$U = U(x)$ – given function ; $U(0) = 0$;

$$\nu = \text{const} ;$$

$$0 \leqslant x \leqslant X \qquad\qquad 0 \leqslant y \leqslant \delta(x);$$

$\delta(x)$ – unknown function to be determined as

part of the solution ;

boundary conditions :

$$.u(x, 0) = v(x, 0) = 0$$

$$u(x, \delta) - U(x)$$

initial conditions :

$$u(0, y) - H_1(y) ; \qquad v(0, y) = H_2(y)$$

Note : No condition for δ appears in the

problem.

Note : Two given functions of x could appear
in the boundary conditions instead of zeros (slip and suction).

1.6.3. Performing of the consecutive operations.

ad I – transformation to the "divergence" form :

$$\left[u(U - u) \right]_x + \left[v(U - u) \right]_y = - U'(U - u) - \frac{\tau_y}{\varrho}$$

$$\tau = \mu u_y$$

ad II – choice of the scheme of approximation :

One strip is chosen bounded by the wall and by the line $\delta(x)$.

Approximation of $u(x, y)$ by a polynomial of – say – 4th degree in y (with coefficients depending on x), satisfying –

– the equation (1.22) at the wall ;

– the boundary conditions of the original problem;

– the postulate of vanishing of the first and second derivatives with respect to y at the "upper" edge of the boundary layer.

These assumptions lead, as it is well-known, to the following approximating polynomial :

(1.24) $$\frac{u}{U} = a(x) \cdot \eta + b(x)\eta^2 + c(x) \cdot \eta^3 + d(x) \cdot \eta^4$$

where

$$(1.25) \begin{cases} \eta = \dfrac{y}{\delta(x)} \ ; \\[2mm] a = 2 + \dfrac{\lambda}{6} \ ; \\[2mm] b = -\dfrac{\lambda}{2} \ ; \\[2mm] c = -2 + \dfrac{\lambda}{2} \ ; \\[2mm] d = 1 - \dfrac{\lambda}{6} \ ; \\[2mm] \lambda = \dfrac{\delta^2}{\nu} U' , \end{cases}$$

the formulae for the coefficients stemming from the four conditions.

ad III – integration with respect to y : .

$$\frac{\partial}{\partial(x)} \int_0^{\delta(x)} \left[u(U-u) \right] dx + v(U-u) \Big|_0^{\delta(x)} = -U' \int_0^{\delta(x)} (U-u) dx - \frac{\tau}{\varrho} \Big|_0^{\delta(x)}$$

After taking account the boundary conditions and
introducing the usual notation of δ_* , δ_{**} and τ_0 the von
Karman's integral formula follows :

$$\frac{d}{dx}(\delta_{**}) + \frac{U'}{U}(2\delta_{**} + \delta_*) = \frac{\tau_0}{\varrho U^2} \tag{1.26}$$

ad IV, V – evaluation of integrals, and differentiation.

The integrals δ_* and δ_{**} can be easily evaluated after substituting of the polynomial (1.24), becoming
functions of λ only. After introducing the integrals as well
as the apropriate expression for τ_0 into (1.26), one obtains
the ordinary differential equation :

$$\frac{d\lambda}{dx} = \frac{L(\lambda; U, U', U''.)}{M(\lambda)} \tag{1.27}$$

where L and M are given functions of their arguments.

ad VI – providing additional equations in order
to "close" the system : does not apply in the present case as

the equation (1.27) contains only one unknown function λ.

ad VII – formulation of suitable conditions.

All the conditions of the original problem are already taken into account. No condition exists as far as $\lambda(x)$ is concerned. The condition can be derived, however, from the equation (1.27) itself, by postulating regularity of the function $\lambda(x)$.

The postulate of the regularity requires simultaneous vanishing of L and M, so that the derivative could remain finite when M = O.

This leads to the known condition :

(1.28) $\lambda = 7.052$

in the stagnation point, where

$$ U = L = O. $$

The value of the derivative in this point corresponding to (1.28) can be obtained by means of the de l'Hospital rule.

1.6.4. Comment : in the present example functions of the unknown functions appeared in the "divergence" form of the partial differential equation, in contrast with previous cases.

1.7. Example of a mixed type equation [3].

1.7.1. Purpose.

– to show application of the MOIR in the case of a mixed type equation, closely related to blunt body problems ;

– to demonstrate a case, when the final system of ordinary differential equations can be solved analytically – in closed form.

1.7.2. Problem.

$$\left.\begin{array}{l} u_y - v_x = 0 \\[2mm] \left[(1-x)\,u\right]_x + v_y = 0 \end{array}\right\} \qquad (1.29)$$

$$u = u(x, y); \qquad v = v(x, y),$$

$$x \geq 0 ; \qquad 0 \leq y \leq 1$$

boundary conditions :

$$u(0, y) = 0 ; \qquad u(x, 1) = x ;$$

$$v(x, 0) = 0 ;$$

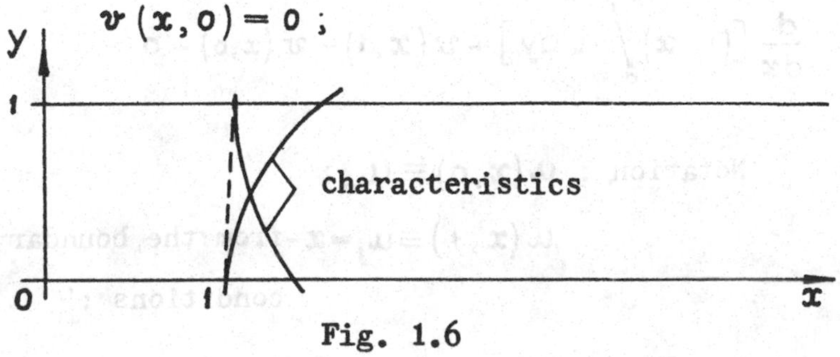

Fig. 1.6

It can be easily seen, that the system (1.29) is of mixed type : elliptic in the region

$$x < 1 ,$$

and hyperbolic in the region

$$x > 1 ,$$

The characteristics to the system (1.29) are represented by two families of parabolas :

(1.30)
$$y = \pm 2 \sqrt{x - 1} + \quad \text{const.}$$

One characteristic of each family is shown in the Fig. 1.6.

1.7.3. Performing of the consecutive operations.

ad I – unecessary ;

ad II – one strip, bounded by $y = 0$; $y = 1$; linear approximation across the strip.

ad III – integration with respect to y :

$$\frac{d}{dx} \int_0^1 v \, dy - u(x,1) + u(x,0) = 0$$

$$\frac{d}{dx} \left[(1-x) \int_0^1 u \, dy \right] + v(x,1) - v(x,0) = 0$$

Notation : $u(x,0) \equiv u_0$;

$\quad\quad u(x,1) \equiv u_1 = x$–from the boundary

$\quad\quad\quad\quad\quad\quad$ conditions ;

$$v(x,0) \equiv v_0 = 0 - \text{ from the boundary}$$

conditions ;

$$v(x,1) \equiv v_1 .$$

Linear approximation :

$$u = u_0 + (u_1 - u_0) y ;$$

$$v = v_1 \cdot y .$$

ad IV - evaluation of the integrals :

$$\int_0^1 u \, dy = \frac{1}{2} (u_1 + u_0) ;$$

$$\int_0^1 v \, dy = \frac{1}{2} v_1 .$$

ad V - differentiation of the evaluated integrals
and substitution into the integral relations :

$$\frac{1}{2} \frac{dv_1}{dx} - u_1 + u_0 = 0 ;$$

$$\frac{1}{2} \frac{d}{dx} \left[(1-x)(u_1 + u_0) \right] + v_1 = 0$$

ad VI - lacking equation : $u_1 = x$.

Final system :

$$\frac{dv_1}{dx} = 2 (x - u_0) ; \tag{1.31}$$

$$(1-x) \frac{du_0}{dx} = u_0 - 2v_1 + 2x - 1 . \tag{1.32}$$

Singularity at $x = 1$.

ad VII - conditions.

$$u_o(0) = 0 \text{ - from the boundary condi-}$$

tions of the original problem;

No condition for v_1

The complementary condition - from the requirement of regularity of $\dfrac{du_o}{dx}$.

The right-hand side of (1.32) must vanish at $x = 1$:

(1.33) $$2v_1 - u_o = 1 \qquad \text{at } x = 1.$$

The solution satisfying both conditions :

(1.34)
$$\begin{cases} u_o = x - 1 + \dfrac{1}{\sqrt{1-x}} \ \dfrac{I_1\left[4\sqrt{1-x}\right]}{I_1(4)} \ ; \\[4mm] v_1 = 2x - \dfrac{3}{2} + \dfrac{I_o\left[4\sqrt{1-x}\right]}{I_1(4)} \ . \end{cases}$$

1.7.4. Exercise : discuss the complementary conditions in the case $N = 2$ using again linear approximation.

1.8. Concluding remarks .

1.8.1. The fundamental concept of the MOIR - getting rid of one of the independent variables of the system of first order partial differential equations by means of integrating the system with respect to this variable.

As the immediate result of such integration a system of so called "integral relations" is obtained.

In case of two independent variables a system of ordinary differential equations follows outright in such a manner. In case of three or more independent variables the integration must be repeated two or more times, with respect to consecutive variables.

1.8.2. The supplementary concept in the MOIR - approximation of the integrals appearing in the integral relations - by means of discrete "values" of the integrands.

"Strips". Linear approximation. Other possibilities. Boundary conditions and their bearing on the most suitable form of approximation (example of the Pohlhausen's polynomial).

1.8.3. The general characteristic feature of the MOIR stems from the two previous points : the method belongs to the so-called "integral-approximation methods" - in contrast with the "derivative-approximation methods" [4]. This feature has a direct bearing on the accuracy of the method, as it is known from numerical analysis.

Further characteristic feature : the method was created for the needs of applied fluid mechanics ; for solving problems as they appear in practice.

It should be well distinguished from methods like selfsimilarity, which also lead to systems of ordinary differential equations, however, only in very special cases of flow, and therefore have rather theoretical significance.

1.8.4. The numerical viewpoint. From the numeric-
al viewpoint it is most convenient, if a problem for a system of
partial differential equations reduces by using the MOIR to the
Cauchy's problem for a system of ordinary differential equations.
Reason : well developed theory and ready numerical procedures
for electronic computers.

If the problem reduces by using the MOIR to a
boundary-value problem or a multipoint problem for a nonlinear
system of ordinary differential equations, there is usually
still some advantage in comparison with finite difference methods
applied directly to the original system of partial differential
equations, especially in cases when only small or medium computers are available.

1.8.5. Restrictions. The MOIR can be applied
exclusively to such a system of partial differential equations,
that every one of them appears in the "divergence" form.

No restrictions are imposed as far as the order
of the system, and its type (hyperbolic, parabolic, mixed) are
concerned.

The number of independent variables as well as
nonlinearity constitute no formal restrictions for applying the
MOIR, however, they have very essential effect on the amount of
preparatory work, necessary for deriving the final system of
ordinary differential equations.

1.8.6. <u>Prospective development of the MOIR</u>. The relatively cumbersome and time-consuming preparatory work connected with applying the MOIR, as well as appearing ever bigger and faster electronic computers shifted lately the main interest to finite – difference methods, which are more straight forward. However, the wellknown difficulties in approximating derivatives by means of finite differences allow one to maintain, that the method of integral relations may still play an important role in specific problems, especially – when only medium computers are available.

("Lucky" applications of the MOIR : one has finally to do with a different mathematical problem, which may turn out to be less restricted from the numerical viewpoint than that for the original system of p a r t i a l differential equations treated by the finite difference method).

References

[1] K. Pohlhausen, Zur näherungsweisen Integration der Differentialgleichung der laminaren Grenzschicht. ZAMM, 1 (1921), 252-268.

[2] А.А. Дородницын, Об одном методе численного решения некоторых нелинейных задач аэрогидро-динамики. Труды III Всесоюзн. матем. съезда 1956, т. III , Изд. АН СССР, 1958, 447-453.

[3] A.A. Dorodnicyn, A contribution to the solution of mixed problems of transonic aerodynamics. Adv. in Aeron. Sci., Vol. II. (Proceedings of the First International Congress in the Aeronautical Sciences Madrid Sept. 1958), Pergamon Press, 1959, 832-844.

[4] F.B. Hildebrand, Finite-Difference Equations and Simulations. Prentice-Hall, inc., Englewood Clifts, New Jersey, 1968.

Chapter 2

APPLICATION TO TYPICAL BLUNT—BODY PROBLEMS

2.1. Motivation for investigating of blunt-body flows .

Hypersonic flows $\longrightarrow$ aerodynamic heating.

The heat streaming into the body moving in the atmosphere (Q), and the heat streaming into the atmosphere (Qg) are functions of the t o t a l drag coefficient c_x and the f r i c t i o n drag coefficient c_{xb}. Very simplified considerations, basing on the Couette flow of viscous, heat conducting ideal gas, and on certain assumptions concerning motion of a body returning from the space into the atmosphere, lead to the following formula :

$$\frac{Q}{Q_g} = \frac{\frac{1}{2}\frac{c_{xb}}{c_x}}{1 - \frac{1}{2}\frac{c_{xb}}{c_x}} , \qquad (2.1)$$

Fig. 2.1

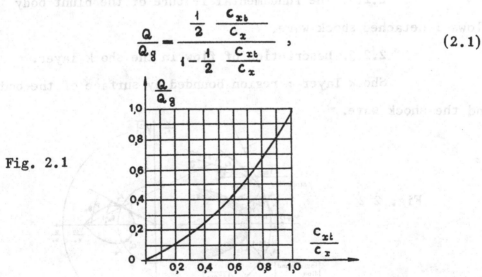

Note : Derivation of the formula (2.1) lies not in the scope of

these lectures, as it involves taking into account viscosity and
heat conductivity of the fluid.

        The obvious advantage of blunt bodies in comparison
with slender, sharp—nosed bodies – as far as the aerodynamic
heating is concerned – follows from the formula (2.1), and from
the Fig. 2.1.

2.2. Some characteristic features of blunt-body flows.

        2.2.1. The frames of reference assumed for descrip-
tion of the flow :

        physically – classical gasdynamics (nonviscous, not
heat conducting gas, obeying the Clapeyron's law) ;

        kinematically – system of reference connected with
the body ——➤ steady flow.

        2.2.2. The fundamental feature of the blunt body
flows : detached shock—wave.

        2.2.3. Description of flow in the shock layer.

        Shock layer : region bounded by surface of the body,
and the shock wave.

Fig. 2.2

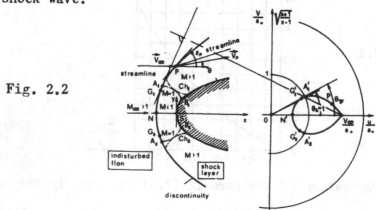

General picture of the flow field in the case of plane flow :

Indisturbed flow in front of the shock wave :

parallel to the x-axis,

homoenergetic, homoentropic, supersonic (Fig. 2.2).

Flow in the shock layer :

homoenergetic ; isoentropic (grad $S \neq 0$, because of the curved

shock) ;

vorticity (rot $\overline{V} \neq 0$ as follows from the Crocco's equation) ;

"mixed" : subsonic-supersonic.

Critical points at the body and at the shock.

Critical lines. Limiting characteristic lines.

Mapping of the flow field onto the hodograph plane (Fig. 2.2.).

Meaning of the symbols in the Fig. 2.2. Points of maximum devia-

tion of a streamline from the direction of the indisturbed flow

(G_1 , G_2).

2.2.4. Equations governing the flow in the shock

layer.

$$\text{div}(\varrho \, \overline{V}) = 0 ; \qquad (2.2)$$

$$\text{grad}\frac{V^2}{2} + \frac{1}{\varrho}\text{grad}\,p = \overline{V} \times \text{rot}\,\overline{V} ; \qquad (2.3)$$

$$\frac{V^2}{2} + \frac{\varkappa}{\varkappa-1}\frac{p}{\varrho} = \frac{V_\infty^2}{2} + \frac{\varkappa}{\varkappa-1}\frac{p_\infty}{\varrho_\infty} ; \qquad (2.4)$$

$$= \frac{V_{max}^2}{2} ;$$

$$= \frac{\varkappa}{\varkappa-1}\frac{p_o}{\varrho_o} .$$

Symbols :

$$V_\infty \;,\; P_\infty \;,\; \varrho_\infty \;-\; \text{velocity, pressure and density of}$$

the indisturbed stream ;

$$\varkappa = \frac{c_p}{c_v} \quad \text{specific heat ratio ;}$$

$$V_{max} \;-\; \text{maximum velocity of the gas ;}$$

$$P_o, \varrho_o - \text{stagnation pressure and density.}$$

2.2.5. Conditions at the body and at the shock wave.

At the body :

(2.5) $$\overline{V} \cdot \overline{n} = 0 \;\;;\;\; \overline{n} \;-\; \text{normal to the body.}$$

At the shock wave – Rankine–Hugoniot conditions
have to be satisfied. Denoting by δ values of gas parameters
(such as velocity, pressure and so on) as follow from the R.-H.-
conditions, we must have :

(2.6) $$p = p_\delta \quad ; \quad \varrho = \varrho_\delta \quad ; \quad u = u_\delta \quad ; \quad v = v_\delta$$

at the shock wave, in the shock layer. As follows from the R.H.
conditions :

(2.7) $$P_\delta = P_\infty \left[1 + \frac{2\varkappa}{\varkappa+1} \left(M_\infty^2 \sin^2 \epsilon - 1 \right) \right]$$

(2.8) $$\varrho_\delta = \varrho_\infty \; \frac{\varkappa + 1}{\varkappa - 1 + \dfrac{2}{M_\infty^2 \sin^2 \epsilon}}$$

$$\frac{u_\delta}{V_\infty} = 1 - \frac{2\sin^2\varepsilon}{\varkappa + 1}\left(1 - \frac{1}{M_\infty^2 \sin^2\varepsilon}\right); \qquad (2.9)$$

$$\frac{v_\delta}{V_\infty} = \frac{\sin 2\varepsilon}{\varkappa + 1}\left(1 - \frac{1}{M_\infty^2 \sin^2\varepsilon}\right), \qquad (2.10)$$

wherein :

ε – angle of the shock wave (Fig. 2.2) ;

u_δ – velocity component in the direction parallel to V_∞ ,

v_δ – velocity component in the direction perpendicular to V_∞ .

2.3. Typical blunt-body problems.

2.3.1. <u>Direct problem</u>.

Given – the body, and data concerning the uniform flow.

Sought – the shock wave, and the flow in the shock layer.

2.3.2. <u>Inverse problem</u>.

Given – the shock wave, and data concerning the uniform flow.

Sought – the body, and the flow in the shock layer.

2.3.3. <u>Pressure problem</u>.

Given – the pressure distribution at the body, and

data concerning the uniform flow.

Sought - the body, the shock, and the flow in the shock layer.

2.3.4. Main difficulties - from the mathematical viewpoint :

- nonlinearity of the system of equations ;

- undefined boundary (either the body or the shock wave not known in advance) ;

- "mixed" system, changing its type (elliptic-hyperbolic) on the critical surface, which is also not known in advance.

2.3.5. Possibilities of simplifications of the system - very limited : to small parts of the flow - field ($\varrho =$ $=$ const), or specific data of the uniform flow (e.g. $M_\infty = \infty$).

2.3.6. The "standard" blunt-body problem of Van Dyke : inverse ; axisymmetric ; $M_\infty = \infty$; paraboloidal shock wave. Suggested in order to form a frame of reference for various methods, conceived for solution of the blunt-body flows [1] .

2.4. Mathematical methods for solving blunt-body problems.

2.4.1. Note : only methods suitable for solution of the flows as described in the frame of assumptions of classical gasdynamics (see par. 2.2.), will be mentioned. Methods basing on physical simplifying assumptions such as constant

density or potentiality of the velocity field-will be disregard-
ed.

No exhaustive survey is intended.

2.4.2. The methods.

2.4.2.1. analytic - seeking the solution
in form of the Taylor series. [2 , 3].
Suited exclusively for inverse problems.
Radius of convergence of the series may be smaller, than the
shock wave distance [4] ⟶ no solution can be obtained in
such cases.

2.4.2.2. methods approximating the system
of partial differential equations (2.2) ÷ (2.4) by a system
of ordinary differential equations :

a) method of integral relations [5] ;

b) "semi" - finite - difference methods [6] ;

c) "semi" - representation by finite series [7, 8].

2.4.2.3. method of finite differences [9,
10].
Concentration in further considerations - exclusively on the
method of integral relations.

**2.5. General equations for plane and axisymmetric flow as obtained by the use of
the MOIR .**

2.5.0. The purpose of this paragraph consists in
derivation of such a system of ordinary differential equations,

which would be suited for solution of <u>all</u> the problems presented
in the par. 2.3.

2.5.1. The case of plane flow.

2.5.1.1. Assumptions

steady flow – in the frame of reference connected with the body
(Fig. 2.3) ;

n, s – frame of reference ;

one strip – first approximation in Dorodnicyn's sense ;

"smooth body – continuous slope; curvature of constant sign
(some of these restrictions will be removed in special cases
treated in the Chapter 3) ; general properties of flow – as
described in the par. 2.2. ;

asymmetric flow will be considered – for the sake of generality.

2.5.1.2. The general scheme and the system of equation.

The <u>general scheme</u> of the flow is shown in the
Fig. 2.3. The curvilinear coordinate s is measured along the
surface of the plane body starting from the stagnation point S.

The stagnation streamline SA is shown to cross
the shock wave at such a point A , wherein the shock is perpendicular. This represents an additional assumption, which will be
discussed and referred to later. Physically – it means, that the
maximum streamline wets the body.

The meaning of some other symbols in the Fig. 2.3
is as follows:

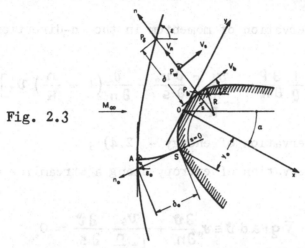

Fig. 2.3

R – local radius of curvature of the body ;

v_b, v_s – velocity components normal and tangential to the body surface, at a point P_w in the shock layer ;

v_b – velocity at the body ;

δ – shock wave distance ;

$_o$ – values corresponding to the stagnation point ;

x, y – a rectangular coordinate system in which the shape of the body and position of the stagnation point are described ;

α – angle of attack.

The system of equations describing the flow in the shock layer stems from the vectorial form presented in the par. 2.2.4., and will be accepted as consisting of the following scalar equations, written in the n,s variables :

continuity –

$$\frac{\partial}{\partial s}(\varrho\, v_s) + \frac{\partial}{\partial n}\left[\varrho\, v_n\left(1 + \frac{n}{R}\right)\right] = 0 \qquad (2.11)$$

conservation of momentum in the n-direction -

$$(2.12) \quad \frac{\partial}{\partial n}\left(\frac{V^2}{2}\right) + \frac{1}{\varrho}\frac{\partial P}{\partial n} + \frac{v_s}{1+\frac{n}{R}}\left[\frac{\partial v_n}{\partial s} - \frac{\partial}{\partial n}\left(1 + \frac{n}{R}\right)v_s\right] = 0 \quad ;$$

conservation of energy - (2.4) ;

conservation of entropy along a streamline -

$$(2.13) \qquad \overline{V}\, \mathrm{grad}\, \vartheta \equiv v_n\frac{\partial \vartheta}{\partial n} + \frac{v_s}{1+\frac{n}{R}}\,\frac{\partial \vartheta}{\partial s} = 0$$

where

$$(2.14) \qquad \vartheta = \frac{P}{\varrho^x}$$

denotes the so-called entropy function.

This system of four equations, three of them being differential equations, contains four functions :

$$v_n \, , \, v_s \, , \, p \, , \, \varrho \, ,$$

and is "closed".

The idea of disregarding the equation of conservation of momentum in the s-direction and closing the system by introducing of the equation (2.13) is due to Bielocerkowski [5]

2.5.1.3. Non-dimensionalising of the system.

Dividing (2.11) by $\varrho_o \, V_{max}$; (2.12) and (2.4) by:

$$\frac{V_{max}^2}{2} = \frac{\varkappa}{\varkappa - 1} \frac{P_o}{\varrho}$$

and finally – the equation (2.13) by :

$$V_{max} \Theta = V_{max} \frac{P_o}{\varrho_o^{\varkappa}}$$

(where P_o, ϱ_o denote stagnation values <u>in front</u> of the shock wave), introducing nondimensional values :

$$\bar{\varrho} = \frac{\varrho}{\varrho_o} \;\;; \quad \bar{p} = \frac{p}{P_o} \;\;; \quad \bar{v}_n = \frac{v_n}{V_{max}} \;\;; \quad v_s = \frac{v_s}{V_{max}} \;\;;$$

$$w = \frac{V}{V_{max}} \;\;; \quad \bar{n} = \frac{n}{b} \;\;; \quad \bar{s} = \frac{s}{b} \;\;; \quad \bar{v} = \frac{v}{\Theta} \;\;,$$

(2.15)

(where b denotes any charateristic length chosen as the scale), one gets after omitting the bars the following system of nondimensional equations :

$$\frac{\partial}{\partial s} (\varrho v_s) + \frac{\partial}{\partial n} \left[\varrho v_n \left(1 + \frac{n}{R} \right) \right] = 0 \;\;; \tag{2.16}$$

$$\frac{\partial}{\partial n} \left(\frac{w^2}{2} \right) + \frac{k}{\varrho} \frac{\partial P}{\partial n} + \frac{v_s}{1 + \frac{n}{R}} \left[\frac{\partial v_n}{\partial s} - \frac{\partial}{\partial n} \left(1 + \frac{n}{R} \right) v_s \right] = 0 \tag{2.17}$$

$$w^2 + \frac{P}{\varrho} = 1 \tag{2.18}$$

$$v_n \frac{\partial v}{\partial n} + \frac{v_s}{1 + \frac{n}{R}} \frac{\partial v}{\partial s} = 0 \tag{2.19}$$

where :

(2.20)
$$k = \frac{\varkappa - 1}{2\varkappa}$$

and

(2.21)
$$w^2 = v_n^2 + v_s^2$$

The boundary conditions for the system were already discussed in the par. 2.2.5. In the notation now accepted, and after taking into account the assumption concerning the maximum entropy streamline; they may be formulated as follows :

At the body :

(2.22) $v_n = 0$ $v = v_b \equiv v_{max}$

where v_{max} is a known value following from the Rankine-Hugoniot conditions :

(2.23) $$v_b = \left[1 + \frac{2\varkappa}{\varkappa + 1} \left(M_\infty^2 - 1 \right) \right] \left[1 + \frac{\varkappa - 1}{\varkappa + 1} \left(M_\infty^2 - 1 \right) \right]^\varkappa M_\infty^{-2\varkappa}$$

At the shock wave :

(2.24)
$$
\begin{cases}
v_s \cos \theta - v_n \sin \theta = u_\delta ; \\
v_s \sin \theta + v_n \cos \theta = v_\delta ; \\
p = p_\delta ; \\
\varrho = \varrho_\delta ;
\end{cases}
$$

where p_δ , ϱ_δ , u_δ , v_δ also follow from the Rankine-Hugoniot conditions, and depend only on the shock wave angle ε as well as on the gasdynamic parameters of the uniform flow (see (2.7)–(2.10)).

2.5.1.4. Derivation of the "divergence" form.

The differential equations of the system i.e. (2.16), (2.17) and (2.19) will be combined in order to obtain the "divergence form" of the system as the precondition for applying the MOIR.

The first "divergence" equation of the system is obtained as combination of (2.17) and (2.16) in the following way :

a) the term :

$$v_n \frac{\partial}{\partial n} (\varrho \, v_n) + \varrho \, v_n^2 + k p) \frac{\partial}{\partial n} \left(1 + \frac{n}{R} \right)$$

is added to and subtracted from (2.17) ;

b) the equation of continuity (2.16) is multiplied by v_n and added to the expression resulting from a) ;

c) after suitable transformation one gets :

$$\frac{\partial}{\partial n} \left[\left(1 + \frac{n}{R} \right) (\varrho \, v_n^2 + k p) \right] + \frac{\partial}{\partial s} (\varrho \, v_n \, v_s) = (\varrho \, v_s^2 + k \, p) \cdot \frac{1}{R} \; , (2.25)$$

the fact, that :

$$R = R(s)$$

being properly taken into account.

Introducing the following notation :

(2.26)
$$
\begin{cases}
H = \varrho\, v_n^2 + k\, p \\[2mm]
g = \varrho\, v_s^2 + k\, p \\[2mm]
z = \varrho\, v_n\, v_s
\end{cases}
$$

one can rewrite (2.25) as follows :

(2.27)
$$
\boxed{\;\frac{\partial z}{\partial s} + \frac{\partial}{\partial n}\left[H\left(1 + \frac{n}{R}\right)\right] = \frac{g}{R}\;.\;}
$$

The second "divergence" equation of the system is obtained from the continuity equation (2.16) taking, however, into account the equation (2.19) representing conservation of entropy along a streamline.

The derivation goes as follows :

a) ousting p from (2.14) and (2.18) one can write:

(2.28)
$$
\varrho = \left(1 - w^2\right)^{\frac{1}{\varkappa - 1}}\, v^{-\frac{1}{\varkappa - 1}}\;;
$$

b) the expression (2.28) is substituted into the continuity equation (2.16), and derivatives are computed as indicated in the resulting equation ;

c) the following symbols are introduced, analogically to (2.26) :

(2.29a)
$$
t = \left(1 - w^2\right)^{\frac{1}{\varkappa - 1}}\, v_s
$$

$$h = \left(1 - w^2\right)^{\frac{1}{\varkappa-1}} v_n \qquad (2.29b)$$

d) introduction of t and h allows to present
the equation as obtained in b) as :

$$v^{-\frac{1}{\varkappa-1}}\left\{\frac{\partial t}{\partial s} + \frac{\partial}{\partial n}\left[h\left(1+\frac{n}{R}\right)\right]\right\} - \frac{1}{\varkappa-1}\left(1-w^2\right)^{\frac{1}{\varkappa-1}} v^{-\frac{1}{\varkappa-1}-1}\left\{v_s\frac{\partial v}{\partial s} + \left(1+\frac{n}{R}\right)v_n\frac{\partial v}{\partial n}\right\} = 0$$

the second member on the left-hand side being obviously naught in
accordance with (2.19) ;

e) finally, therefore :

$$\boxed{\frac{\partial t}{\partial s} + \frac{\partial}{\partial n}\left[h\left(1+\frac{n}{R}\right)\right] = 0} \qquad (2.30)$$

which represents the desired second "divergence" equation of the
system.

Comment : The continuity equation (2.16) appears already in the
divergence from, and it could be left as it is from the viewpoint
of application of the MOIR. However, because of the transformations described the equation (2.30) includes more information
than the equation (2.16), namely the information of the entropy
conservation (2.19).

The equation (2.19) will be not used any more in
further considerations.

Of course, the two "divergence" equations (2.27)

and (2.30) are not equivalent to the system (2.16) – (2.19).

2.5.1.5. Derivation of the system of ordinary differential equations.

Both equations (2.27) and (2.30) will be treated now by means of the MOIR, as they appear already in the necessary "divergent" form.

Integrating them with respect to n within the bounds $0 \div \delta(s)$ in accordance with the accepted "scheme of approximation", one gets :

$$(2.31) \quad \begin{cases} \dfrac{d}{ds} \displaystyle\int_0^\delta t \, dn - \dfrac{d\delta}{ds} t_\delta + h_\delta\left(1 + \dfrac{\delta}{R}\right) = 0 \, ; \\[3mm] \dfrac{d}{ds} \displaystyle\int_0^\delta z \, dn - \dfrac{d\delta}{ds} z_\delta + H_\delta\left(1 + \dfrac{\delta}{R}\right) - H_b = \dfrac{1}{R} \displaystyle\int_0^\delta g \, dn \, . \end{cases}$$

The symbol $_b$ refers to the body surface, $_\delta$ – to the shock wave. The value :

$$h_b = 0$$

is already taken into account in (2.31).

Evaluating the integrals in (2.31) when treating g, t, z as linear functions of n, one obtains :

$$(2.32) \quad \begin{cases} \delta \dfrac{dz_\delta}{ds} - z_\delta \dfrac{d\delta}{ds} = \dfrac{\delta}{R}\left(g_\delta + g_b\right) - 2\left[\left(1 + \dfrac{\delta}{R}\right) H_\delta - H_b\right] \, ; \\[3mm] \dfrac{dt_\delta}{ds} + \dfrac{dt_b}{ds} + \dfrac{t_b - t_\delta}{\delta} \dfrac{d\delta}{ds} = -2\left(\dfrac{1}{\delta} + \dfrac{1}{R}\right) h_\delta \, . \end{cases}$$

It can be easily seen, that all the functions denoted $_\delta$ depend only on ε and some of them on θ . It follows from the conditions (2.24) and from the definition of

$$H_\delta \ , \ g_{d\delta} \ , \ z_\delta \ , \ t_\delta \ , \ h_\delta \qquad\qquad (2.33)$$

(comp. (2.26), (2.29)). Therefore we can write :

$$f_\delta = f_\delta (\varepsilon , \theta) \qquad\qquad (2.34)$$

denoting any of the functions (2.33).

On the other hand, all the functions (2.26) and (2.28) at the body, i.e. denoted $_b$ depend on v_b only.

It follows from the fact, that the entropy function at the body is constant, and equal*) its maximum value (2.23). Indeed, taking into account (2.22), one has :

$$\varrho_b = \left(1 - v_b^2 \right)^{\frac{1}{\varkappa - 1}} \ v_b^{\div \frac{1}{\varkappa - 1}} \qquad\qquad (2.35)$$

accordingly to (2.28), and

$$p_b = \left(1 - v_b^2 \right) \varrho_b \qquad\qquad (2.36)$$

---

*) This is obvious in the symmetric case. In the asymmetric case this property must be postulated. Comp. the discussion in [15] and [16] .

accordingly to (2.18). Further, by virtue of (2.26) and (2.28) :

$$(2.37) \quad \begin{cases} H_b = k\, p_b = H_b(v_b) \\[2mm] g_b = \varrho_b\, v_b^3 + k\, p_b = g_b(v_b) \\[2mm] z_b = 0 \\[2mm] t_b = \left(1 - v_b^2\right)^{\frac{1}{\varkappa - 1}} v_b \\[2mm] h_b = 0 \ . \end{cases}$$

Denoting by f_b any of the functions (2.37) we can write in analogy with (2.34) :

$$(2.38) \qquad\qquad f_b = f_b(v_b)$$

<u>Summarising</u> : we obtained so far the system (2.32) of two ordinary differential equations, containing four functions :

$$(2.39) \qquad\qquad v_b, \varepsilon, v, \delta \ .$$

Which ones of them are given, and which are sought depends on the nature of the problem, as will be shown later.

<u>Geometric considerations</u>. Because the strip, that we have chosen in our scheme of approximation, is bounded by the curved line $\delta(s)$ which is-generally - not specified, we have to provide an additional differential equation, as it was

stated in Chapter 1.

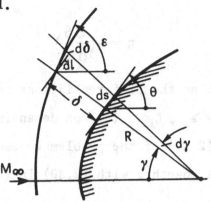

Fig. 2.4.

The equation sought follows from purely geometric considerations (Fig. 2.4). There is :

$$d s = R d \gamma$$

$$d \ell = (R + \delta) d \gamma$$

$$\frac{d \delta}{d \ell} = tg (\varepsilon - \theta)$$

wherefrom the <u>third</u> equation of the system can be easily obtained, as :

$$\frac{d \delta}{d s} = \left(1 + \frac{\delta}{R}\right) tg (\varepsilon - \theta) . \qquad (2.40)$$

Taking into account, that :

$$\gamma + \theta = \frac{\pi}{2}$$

or

$$d \gamma = - d \theta$$

we obtain also :

(2.41)
$$R = -\left(\frac{d\theta}{ds}\right)^{-1}$$

Performing the differentiations in (2.32) of the auxiliary functions z_δ , t_δ , t_b as depending on the proper unknown functions (2.39) of the problem we can finally write these two equations together with (2.40) in the following convenient form :

$$\tilde{A}\frac{d\varepsilon}{ds} + \tilde{B}\frac{d\upsilon_b}{ds} + \tilde{C}\frac{d\delta}{ds} + \tilde{D}_2\delta\frac{d\theta}{ds} = \tilde{D}_1 \; ;$$

(2.42)
$$\tilde{E}\frac{d\varepsilon}{ds} \qquad\qquad + \tilde{F}\frac{d\delta}{ds} + \tilde{G}_2\delta\frac{d\theta}{ds} = \tilde{G}_1 \; ;$$

$$\frac{d\delta}{ds} + L_1\delta\frac{d\theta}{ds} = L_1 \; ;$$

It can be seen now, that one of the functions ε , $\upsilon_b, \delta, \theta$ must be given, in order to "close" the system.

The system (2.4.2) is nonlinear, and its coefficients are represented by rather complicated expressions. Denoting - for the sake of abbreviation :

$$a_z = \sqrt{\frac{\varkappa - 1}{\varkappa + 1}}$$

$$\tau_\delta = \left(1 - w_\delta^2\right)^{\frac{1}{\varkappa - 1}}$$

$$\tau_b = (1 - v_b^2)^{\frac{1}{\varkappa - 1}}$$

$$P = 1 - v_{s\delta}^2 / a_x^2 - v_{n\delta}^2$$

$$Q = 2 \, v_{s\delta} \, v_{n\delta} / (\varkappa - 1)$$

$$U_1 = \frac{\varkappa - 1}{2} \, \sin^2 \varepsilon + \frac{1}{M_\infty^2}$$

$$U_2 = \cos 2\varepsilon + \frac{1}{M_\infty^2 \sin^2 \varepsilon}$$

we can represent the coefficients by means of the following formulae :

$$\tilde{A} = - \frac{2 \, w_\infty \, \tau_\delta^{2 - \varkappa}}{\varkappa + 1} \left[(P \sin \theta - Q \cos \theta) U_s - (P \cos \theta + Q \cos \theta) \sin 2\varepsilon \right]$$

$$\tilde{B} = \tau_b^{2 - \varkappa} (1 - v_b^2 / a_x^2)$$

$$\tilde{C} = \frac{1}{\varrho} (\tau_b v_b - \tau_\delta v_{s\delta})$$

$$\tilde{D}_1 = - \frac{2 \, \tau_\delta \, v_{n\delta}}{\delta} \qquad\qquad (2.43a)$$

$$\tilde{D}_2 = \frac{\upsilon_\delta \upsilon_{n\delta}}{\delta} = \frac{\tilde{D}_1}{2}$$

$$\tilde{E} = \delta \left\{ \frac{\varkappa+1}{2} \upsilon_{n\delta} \upsilon_{s\delta} \frac{\varrho_\infty}{M_\infty^2} \frac{\sin 2\varepsilon}{U_1^2} + \right.$$

$$+ \varrho_\infty \frac{2w^2}{\varkappa+1} \left[(\upsilon_{s\delta} \cos\Theta + \upsilon_{n\delta} \sin\Theta) U_2 + \right.$$

$$+ \left. \left. (\upsilon_{s\delta} \sin\Theta - \upsilon_{n\delta} \cos\Theta) \sin 2\varepsilon \right] \right\}$$

$$\tilde{F} = -\varrho_\delta \upsilon_{s\delta} \upsilon_{n\delta}$$

$$\tilde{G}_1 = -2\varrho_\delta \left[\upsilon_{n\delta}^2 + k(1-w_\delta^2) \right] + 2k\varrho_b(1-\upsilon_b^2)$$

$$\tilde{G}_2 = -\varrho_\delta \left[\upsilon_{n\delta}^2 + k(1-w_\delta^2) \right] + k\varrho_b(1+\upsilon_b^2/a_x^2) \, ;$$

(2.43b)
$$L_1 = tg(\varepsilon - \Theta) \, .$$

Comment : The amount of work necessary for treating a real problem by the MOIR is probably evident to the audience - at this point.

2.5.2. The case of axisymmetric flow.

2.5.2.1. The "divergence" form of the equations.

In the axisymmetric case the equations of continuity and of conservation of momentum in the n-direction, can
be brought by means of transformations analogical to that presented in the previous paragraph, to the following "divergence"
form :

$$\frac{\partial(\nu r)}{\partial s} + \frac{\partial}{\partial n}\left[hr\left(1 + \frac{n}{R}\right)\right] = 0 \qquad (2.44)$$

$$\frac{\partial(\varkappa r)}{\partial s} + \frac{\partial}{\partial n}\left[Hr\left(1 + \frac{n}{R}\right)\right] = \frac{gr}{R} + kp\left(1 + \frac{n}{R}\right)\cos\Theta \quad (2.45)$$

the equation (2.40) remaining unchanged.

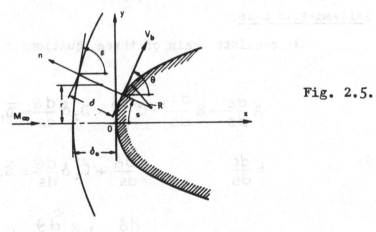

Fig. 2.5.

The new symbol denotes (Fig. 2.5):

$$r = y + n\cos\Theta\;;$$

where

$$y = y(s)$$

is a coordinate of a point of the body surface.

2.5.2.2. <u>Evaluation of the integrals</u>.

After integration of (2.44) and (2.45) with respect to n, in the bounds $0 \div \delta(s)$ one is faced with the problem of approximating the integrands. Chosing linear approximation of t, z, g (and <u>not</u> that of tr, gr, zr — which some authors did), and assuming additionally :

$$p = \frac{n}{\delta}(p_\delta - p_b) + p_b \, ,$$

one arrives finally at the system of equations quite similar to (2.4.2), however, with different coefficients.

2.5.2.3. <u>The final system of equations in the axisymmetric case</u>.

It consists again of three equations :

$$\bar{A}\frac{d\varepsilon}{ds} + \bar{B}\frac{d\upsilon_b}{ds} + \bar{C}\frac{d\delta}{ds} + D_2 \delta \frac{d\theta}{ds} = \bar{D}_1 ;$$

(2.46)

$$E\frac{d\varepsilon}{ds} \qquad\qquad + \bar{F}\frac{d\delta}{ds} + G_2 \delta \frac{d\theta}{ds} = \bar{G}_1 ;$$

$$\frac{d\delta}{ds} + L_1 \delta \frac{d\theta}{ds} = L_2 ,$$

with the following coefficients :

(2.47a)

$$\bar{A} = \tilde{A}\left(y + \frac{2}{3}\delta\cos\theta\right)$$

$$\bar{B} = \tilde{B}\left(y + \frac{1}{3}\delta\cos\theta\right)$$

$$\bar{C} = \tilde{C}\left(y + \frac{2}{3}\delta\cos\theta\right);$$

$$\bar{D}_1 = \tilde{D}_1\left(y + \delta\cos\theta\right) + \sin\vartheta\left(\tau_\delta v_{s\delta} + \tau_b v_b\right);$$

$$\bar{D}_2 = \tilde{D}_2\left(y + \delta\cos\theta\right) + \frac{\sin\theta}{3}\left(\tau_\delta v_{s\delta} + \tau_b v_b\right) +$$

$$+ \frac{\tau_\delta}{3}\left(v_{s\delta}\sin\theta + v_{n\delta}\cos\theta\right);$$

$$\bar{E} = \tilde{E}\left(y + \frac{2}{3}\delta\cos\theta\right)$$

$$\bar{F} = \tilde{F}\left(y + \frac{2}{3}\delta\cos\theta\right)$$

$$\bar{G}_1 = \tilde{G}_1\left(y + \delta\cos\theta\right) - \delta\sin\theta\,\tilde{F} -$$

$$- \delta\cos\theta\,\frac{\varkappa - 1}{2\varkappa}\left[\left(1 - w_\delta^2\right)\varrho_\delta - \left(1 - v_b^2\right)\varrho_b\right] +$$

$$\bar{G}_2 = \tilde{G}_2\left(y + \cos\theta - \frac{2}{3}\delta\sin\theta\,\tilde{F} -\right.$$

$$- \frac{\delta}{3}\cos\theta\,\frac{\varkappa - 1}{2\varkappa}\left[\left(1 - w_\delta^2\right)\varrho_\delta - \left(1 - v_b^2\right)\varrho_b\right] +$$

$$+ \cos\theta\,\frac{\delta}{3}\left(+2\varrho_b v_b^2 + v_{n\delta}^2\varrho_\delta\right)$$

$$(2.47\text{b})$$

it is to be understood, that A, B... are expressed by (2.43).

2.5.2.4. The singularity at the stagnation point.

It can be seen, that the equations (2.46) do not determine the derivatives at the stagnation point :

$$s = y = v_b - v_{sb} = \cos\theta = \cos\varepsilon = \frac{d\delta}{ds} = 0,$$

which was not the case in the plane flow, and which is caused by y appearing now in the formulas for the coefficients.

This singularity is easy to get rid of by differentiation of the first two equations (2.46) with respect to s , and substituting then, to the expressions obtained, the values of s, y, v_b and so on corresponding to the stagnation point.

If e.g. the radius R is a given function, the following system can be obtained :

$$a_1 \frac{d\varepsilon}{ds} + b_1 \frac{dv_b}{ds} = c_1 \; ;$$

$$a_2 \frac{d\varepsilon}{ds} \qquad\qquad = c_2 \; ,$$

$a_1, a_2, \ldots c_2$ being known constants.

2.5.3. Final remark.

The final system (2.42) of three ordinary differential equations for the case of plane flows differs from that for axisymmetric flows (2.46) only in the coefficients.

Therefore we shall write both systems in the same form :

(2.48a) $$+ B \frac{dv_b}{ds} + C \frac{d\delta}{ds} + D_2 \delta \frac{d\theta}{ds} = D_1$$

$$E \frac{d\varepsilon}{ds} \qquad\qquad + F \frac{d\delta}{ds} + G_2 \delta \frac{d\theta}{ds} = G_1 \ ;$$

$$\frac{d\delta}{ds} + L_1 \delta \frac{d\theta}{ds} = L_1 \ , \qquad\qquad (2.48b)$$

and it is only to be understood, that the coefficients A, B,...
are determined either by (2.43) or (2.47), depending whether
the plane, or the axisymmetric flow is considered.

The system (2.48) is, as it was our intention,
suitable for dealing with all the problems named in the par.
2.3.

2.6. The direct problem.

2.6.1. System of equations.

In the direct problem the shape of the body is
given, and therefore

$$\theta = \theta(s) \qquad ; \qquad y = y(s) \qquad\qquad (2.49)$$

are known functions, so that in the system (2.48)

$$\varepsilon = \varepsilon(s); \quad \vartheta_b = \vartheta_b(s); \quad \delta = \delta(s) \qquad\qquad (2.50)$$

are unknown.

Introducing the following symbols :

$$G = G_1 - G_2 \delta \frac{d\theta}{ds} \qquad\qquad (2.51a)$$

$$D = D_1 - D_2 \delta \frac{d\theta}{ds}$$

(2.51b) $$L = L_1 - L_1 \delta \frac{d\theta}{ds}$$

we can represent the system (2.48) in the Cauchy's normal form :

(2.52) $$\frac{d\delta}{ds} = L$$

(2.53) $$\frac{d\varepsilon}{ds} = \frac{G - FL}{E}$$

(2.54) $$\frac{d v_b}{ds} = \frac{1}{B}\left[D - CL - \frac{A}{E}(C - FL)\right]$$

the right-hand sides depending only on s and on the three
unknown functions (2.50).

2.6.2. Conditions.

Note : In the whole chapter 2. only symmetric problems will be
considered, as they are "typical" - in the frame of the method
of integral relations.

At the stagnation point :

(2.55)
$$\begin{cases} v_b = 0 \ ; \\[2mm] \varepsilon = \frac{\pi}{2} \ ; \\[2mm] \delta = \delta_0 \ - \text{unknown.} \end{cases}$$

Another condition must obviously be included, making up for the one unknown initial value δ_o. It follows from the equation (2.54). At

$$v_b = a_*$$ (2.56)

the denominator B vanishes – see (2.43). Therefore the numerator must also vanish at this point :

$$D - CL - \frac{A}{E}(C - FL)\bigg|_{v_b - a_*} = 0 .$$ (2.57)

The requirement (2.57) represents the lacking complementary condition.

2.6.3. Solution technique.

The mathematical problem described by the system (2.52) – (2.54) of three ordinary differential equations with the initial conditions (2.55), and the complementary condition (2.57) represents a two point problem, which has to be solved numerically by consecutive approximations. The main difficulty of the solution is due to the fact, that the point S_* at which the complementary condition (2.57) has to be satisfied, is unknown in advance. Also the type of singularity at this point (saddle) causes severe numerical instability.

Nevertheless the problem was solved successfully, and a very simplified flow chart representing the technique of computing the consecutive approximations of δ_o is shown in the

Fig. 2.7.

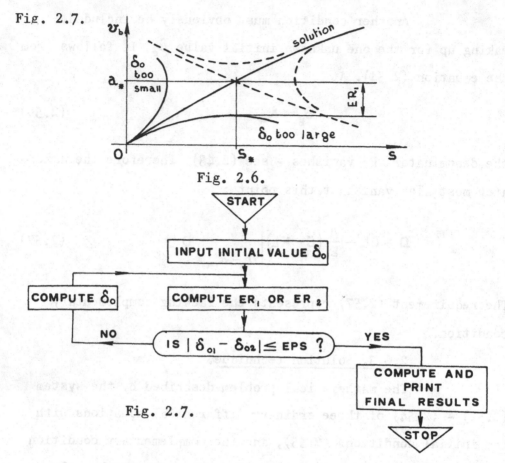

Fig. 2.6.

Fig. 2.7.

Definition of the "error indicator" ER_1 can be taken directly from the Fig. 2.6. The "error indicator" ER_2, corresponding to "too small" initial values δ_o of the shock stand-off distance (see Fig. 2.6) is taken straight from the complementary condition (2.57), as :

$$ER_2 = D - CL - \frac{A}{B}(C - FL)\Big|_{v_b = a_*}$$

The values of δ_o denoted δ_{o1} and δ_{o2} in the

Fig. 2.7. correspond to two different situations (Fig. 2.6), when either ER_1 or ER_2 is computed.

Computation of ER_1 or ER_2 is done in course of numerical integration of the system (2.52) - (2.54), starting from the stagnation point every time.

2.6.4. Results and their confirmation by experiment.

Some results taken from Bielocerkowski's papers [5] are represented in the Figs. 2.8. ÷ 2.11.

Consistency with experiments turns out to be surprisingly good even for $N = 1$, however, only at the sufficiently large Mach Numbers M_∞. At smaller Mach Numbers $M_\infty < 2$ more strips are necessary.

It should be pointed out, that in the vicinity of $M_\infty = 1$ the method fails.

It can fail also for certain shapes of the body. Reasons of such "failures" were investigated by Luczywek [14] - see also [13] . Being not able to discuss these reasons in detail in the frame of the present course, we will confine ourselves to the following remark.

Fig. 2.8

Fig. 2.9

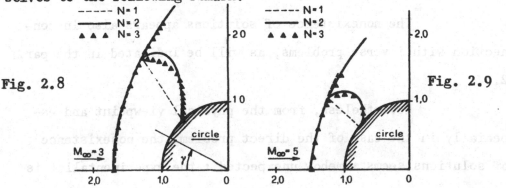

From the purely mathematical viewpoint there is nothing strange in the fact, that the problem with the initial and complementary conditions posed for the system (2.48) may have no solution. The system is nonlinear, with very complicated coefficients, and the existence of a solution to such the problem has never been formally proved. Therefore it is not surprising from the formal viewpoint, that for some functions $\Theta(s)$ as well as some values of M_∞ and $\varkappa$ the solution may not exist.

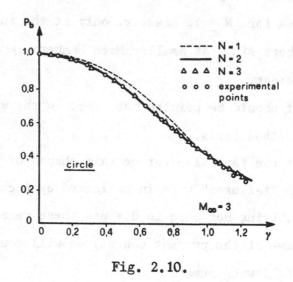

Fig. 2.10.

The nonexistence of solutions appears also in connection with inverse problems, as will be indicated in the par. 2.8.

Nevertheless, from the physical viewpoint and especially in the case of the direct problem, the nonexistence of solutions seems somehow unexpected : the experimentalist is

only too accustomed to the simple fact, that a shock layer
always forms around a blunt body at supersonic speeds, regard -
less of the details of the body shape or of the value of the
Mach Number. This is quite true, we must remember, however,
that we deal not with real bodies and flows, but with equations.

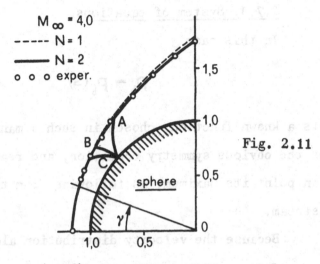

Fig. 2.11

The more, our system (2.48) of ordinary differential equations
has been obtained by the use of rather crude formal assumptions
of the linearity of some integrands, and it may represent - cor-
respondingly - very crude approximation of the original system
of the partial differential equations of classical gasdynamics,
and very crude description of the flow under consideration.

The distortion of the physical description of the
flow field due to those formal assumptions - is not clear in
advance, have been not analysed, and no wonder, that it may
(and does) appear significant in case of certain body shapes
resulting in the said nonexistence of the solutions.

Application of the original system of equations usually widens, however, the range of existence of the solutions.

2.7. The pressure problem.

2.7.1. System of equations.

In this case

(2.58)
$$P_b = P_b(s)$$

represents a known function, chosen in such a manner, that it satisfies the obvious symmetry condition, and reaches at the stagnation point its maximum, as following from the data for the uniform stream.

Because the velocity distribution along the body :

$$v_b = v_b(s)$$

as well as its derivative

$$\frac{dv_b}{ds} = v_b'(s)$$

follow immediately from (2.58), (see (2.36) and (2.35)), the system (2.48) can be written in this case in the following form :

(2.59)
$$\frac{d\theta}{ds} = \frac{D_1 - CL_1 - \frac{A}{E}(G_1 - FL_1) - Bv_b'}{\delta\left[D_2 - CL_1 - \frac{A}{E}(G_2 - FL_1)\right]}$$

$$\frac{d\varepsilon}{ds} = \frac{1}{E}\left[G_1 - FL_1 - (G_2 - FL_1)\delta\frac{d\theta}{ds}\right]; \qquad (2.60)$$

$$\frac{d\delta}{ds} = L_1\left(1 - \delta\frac{d\theta}{ds}\right). \qquad (2.61)$$

It is to be understood, that the symbol $\frac{d\theta}{ds}$ in

the equations (2.60) and (2.61) denotes in fact the right-hand

side of the equation (2.59). Therefore, at least in the case

of plane flow, the right-hand sides of all the three equations

of the system (2.59) - (2.61) depend only on the independent

variable s , and on the unknown functions $\theta, \varepsilon, \delta$, so that

the system is closed.

In the axisymmetric case the system is not closed,

because the unknown function

$$y = y(s)$$

appears in the coefficients. Therefore two complementary equa-

tions

$$\frac{dy}{ds} = \sin\theta; \qquad (2.62)$$

$$\frac{dx}{ds} = \cos\theta; \qquad (2.63)$$

must be included into the system, x and y denoting coordinates

of the body (see Fig. 2.5). The same two equations are usually

included also in the plane case, because it is more convenient to obtain the body already in the rectangular coordinates.

2.7.2. Conditions.

As previously, the main conditions refer to the stagnation point:

$$s = 0,$$

and they have, in the both cases considered (plane as well as axisymmetric), the form of initial conditions :

$$(2.64) \quad \begin{cases} \theta(0) = \dfrac{\pi}{2} \ ; \\[2mm] \varepsilon(0) = \dfrac{\pi}{2} \ ; \\[2mm] \delta(0) = \delta_0 \ - \text{unknown constant} \ ; \\[2mm] y(0) = 0 \ ; \\[2mm] x(0) = 0 \ . \end{cases}$$

The complementary condition follows from the equation (2.59), and it requires, that the numerator and denominator of the right-hand side of this equation vanish "simultaneously".

2.7.3. Solution technique.

– as in 2.6.3.

2.7.4. Results and their consistency with experiments.

As far as we know, the only results have been obtained by ourselves [12] for the case of plane flow, and by Kentzer [11] for axisymmetric flow.

No suitable experiments have been performed, to our knowledge.

Some of our results are represented in the Figs. 2.12 and 2.13, the first one of them showing the accepted velocity distributions, the second one - the corresponding bodies and shock waves.

A simple formula, relating the radius of curvature, the initial shock wave distance and the velocity gradient at the stagnation point was also obtained [12] :

$$\frac{1}{\left(\frac{R_o}{\delta_o}\right)} + 2 = \frac{1}{T_o}\left[\frac{d\nu_b}{d(s/\delta_o)}\right]_o \qquad (2.65)$$

where T_o denotes a known positive function of Mach Number and specific heat ratio :

$$T_o = T_o(M_\infty ; \varkappa) > 0 \qquad (2.66)$$

which has the following value for $M_\infty = 3$ and $\varkappa = 1.4$:

$$T_o(3 ; 1.4) = 0.0984. \qquad (2.67)$$

2.8. The inverse problem .

2.8.1. System of equations.

If the shock wave is given - by means of a function:

$$x = x(y) \qquad (2.68)$$

the system (2.48) must be suitably transformed before using it for solving the problem. Because y appears now to be the most

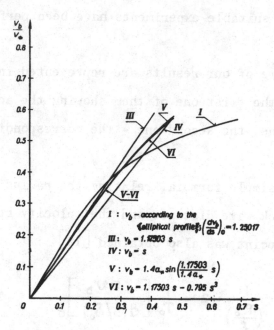

$$I : v_b - \text{according to the}$$
$$\left\{\text{elliptical profile}\right\}; \left(\frac{dv_b}{ds}\right)_0 = 1.25017$$
$$III : v_b = 1.17503 \, s$$
$$IV : v_b = s$$
$$V : v_b = 1.4 a_* \sin\left(\frac{1.17503}{1.4 a_*} s\right)$$
$$VI : v_b = 1.17503 \, s - 0.795 \, s^3$$

Fig. 2.12

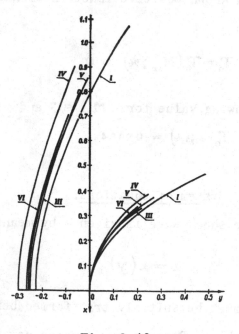

Fig. 2.13

convenient independent variable, we will multiply (2.48) by $\dfrac{ds}{dy}$
including then a further equation containing this de-
rivative.

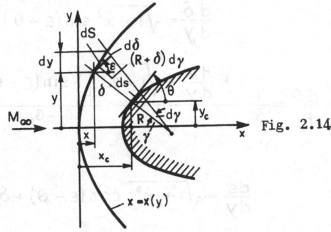

Fig. 2.14

This equation follows easily from geometric relations represented in the Fig. 2.14. We have, namely :

$$dS^2 = \left[1 + \left(\frac{dx}{dy} \right)^2 \right] dy^2 ;$$

$$ds^2 = d\delta^2 + (R + \delta)^2 d\gamma^2 ;$$

$$Rd\gamma = ds ;$$

$$R = - \frac{ds}{d\theta} ;$$

wherefrom :

$$1 + \left(\frac{dx}{dy} \right)^2 = \left(\frac{d\delta}{dy} \right)^2 + \left(\frac{ds}{dy} - \delta \frac{d\theta}{dy} \right)^2. \qquad (2.69)$$

After suitable transformations of the system (2.48)

extended now by the equation (2.69) we obtain, as usually, the
Cauchy's normal form :

(2.70)
$$\frac{d\delta}{dy} = \sqrt{1 + \dot{x}^2} \, \sin(\varepsilon - \theta)$$

(2.71)
$$\frac{d\theta}{dy} = \frac{1}{\delta} \, \frac{E \frac{d\varepsilon}{dy} + \sqrt{1 + \dot{x}^2} \left[F \sin(\varepsilon - \theta) - G_1 \cos(\varepsilon - \theta) \right]}{G_1 - G_2}$$

(2.72)
$$\frac{ds}{dy} = \sqrt{1 + \dot{x}^2} \, \cos(\varepsilon - \theta) + \delta \frac{d\theta}{dy}$$

(2.73)
$$\frac{d\upsilon_b}{dy} = \frac{1}{B} \left[D - CL - \frac{A}{E}(G - FL) \right] \cdot \frac{ds}{dy} \, .$$

As previously, we include two further equations :

(2.74)
$$\frac{dy_c}{dy} = \frac{ds}{dy} \cdot \sin\theta \, ;$$

(2.75)
$$\frac{dx_c}{dy} = \frac{ds}{dy} \cdot \cos\theta \, ,$$

expressing coordinates of a point at the body surface.

In the system (2.70) ÷ (2.75) the $\dot{x}$ denotes de-
rivative of (2.88) ; of course ε and $\frac{d\varepsilon}{dy}$ are known functions
of this derivative, and can be expressed as follows :

(2.76a)
$$\varepsilon = \text{arctg} \, \frac{1}{\dot{x}}$$

$$\frac{d\varepsilon}{dy} = - \frac{\dot{x}}{1 + \dot{x}^2} \qquad (2.76b)$$

2.8.2. Conditions.

In the stagnation point $(s = 0)$:

$$
\left.
\begin{aligned}
\delta\,(0) &= \delta_0 \ - \text{unknown constant} \\
\theta\,(0) &= \frac{\pi}{2}\ ; \\
s\,(0) &= 0\ ; \\
v_b(0) &= 0\ ; \\
x_c\,(0) &= 0\ ; \\
y_c\,(0) &= 0\ .
\end{aligned}
\right\} \qquad (2.77)
$$

The complementary condition follows from the equation (2.73), and has exactly the same form (2.57), as in the case of the direct problem.

2.8.3. Results.

It has been shown [13] , that the MOIR - in its first approximation - can be successfully applied to inverse blunt-body problems, although not for all shapes of the shock wave. (E.g. the Van Dyke's "standard" problem could not be solved).

(The physical reason of this result is obvious and wellknown).

We proved (see [13]), that if the solution to the inverse problem exists, it lies within the bounds :

$$(2.78) \qquad \frac{1}{2} \leqslant \frac{\delta_o}{\delta_{o_{max}}} < 1 \, ,$$

where

$$(2.79) \qquad \frac{\delta_{o_{max}}}{R} = \frac{1}{\varkappa - 1} \left\{ \varkappa^2 + 4\varkappa - 1 - \left[\frac{(\varkappa - 1)^{\varkappa + 1}}{4\varkappa} \right]^{\frac{1}{\varkappa - 1}} \right\} \, ,$$

for $M_\infty = \infty$.

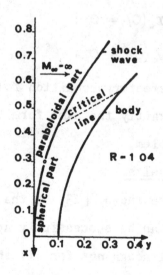

Fig. 2.15

Just in order to give some idea, what the solution looks like, the Fig. 2.15 is presented, taken from [13] . It concerns an

axisymmetric case, the shock wave being assumed as consisting
partly of a sphere, and partly of the paraboloid :

$$y = \frac{1}{2} x^2 \qquad (2.80)$$

the radius of the sphere being indicated in the Fig. 2.15.

References

[1] M.D. Van Dyke, Hypersonic flow behind a paraboloidal shock wave, J. Méc. Appl., 4, 4 (1965), 476-495.

[2] C.C. Lin and S.I. Rubinow, On the Flow Behind Curved Shocks, J. Math. and Phys., Vol. 27, No. 2 July 1948, 105-129.

[3] H. Cabannes, Contribution à l'étude théorique des fluides compressibles, Ecoulements transsoniques, Ondes de choc. Chapitre III, Etude de l'onde de choc détachée au voisinage de son sommet. Ecole Normale Supérieure, Annales Scientifiques, Ser. 3, Vol. 69, 1952, 31-46.

H. Cabannes, Tables pour détermination des ondes de choc détachées, La Recherche Aéronautique, No. 49, Jan-Feb. 1956, 11-15.

[4] M.D. Van Dyke, A Model of Supersonic Flow Past Blunt Axisymmetric Bodies, with Application to Chester's Solution. Journ. Fluid Mech., Vol. 3., Feb. 1958, 515-522.

[5] О.М. Белоцерковский, Выч. Мат., 3,(1958),149-185. Прикл.Мат., Mex. 2, 22 (1958) , 206-219 Прикл.Мат., Mex. 3, 24 (1960) .

[6] M. Inouye, J.V. Rakich, H. Lomax, A description of numerical mathods and computer programs for two-dimensional and axisymmetric supersonic flow over bunt-nosed and flared bodies, NASA TN D - 270, August.

[7]   R.J. Swigart, A Theory of Asymmetric Hypersonic Blunt
        Body Flows, AIAA Journal, 1, 5 (1963), 1034–1046.

[8]   А.П. Базжин, А.А. Гладков, К решению обратной
        задачи методом разложения в ряды, Инж.
        Журнал, 3, 111   (1963) 517–518.

[9]   I.O. Bohachevsky, E.L. Rubin, R.E. Mates, A Direct Method
        for Computation of Nonequilibrium Flows with
        Detached Shock Waves, Abstract, AIAA 2nd Aerospace
        Sciences Meeting, New York, Jan. 25–27, 1965 (AIAA
        Paper No. 65–24).

[10]   V.V. Rusanow, A three-dimensional supersonic gas flow
        past smooth blunt bodies, Proc. of the 11th Intern.
        Congress of Appl. Mech., Munich 1964, 774–778.

[11]   C.P. Kentzer, The inverse blunt-body problem, Preprint,
        XV International Astronautical Congress, Warsawa,
        Sept. 7–12, 1964.

[12]   W.J. Prosnak, E. Luczywek, On the inverse asymmetric
        hypersonic blunt-body problem, Fluid Dynamics
        Transactions 3, (1967).

[13]   J.M. Breiter, E.Luczywek, Inverse blunt-body problem by
        the method of integral relations, Fluid Dynamics
        Trans., 4, 155–162.

[14]   E. Luczywek, Analysis of the method of integral relations
        in application to investigation of blunt-body flows
        (Diss., Warsaw Technological University, Warsaw
        1966 – in Polish).

[15]   W.J. Prosnak, The asymmetric hypersonic blunt-body prob-
         lem, Fluid Dynamics Transactions, vol. 2, (1965),
         457-476.

[16]   R.J. Swigart, The direct asymmetric hypersonic blunt-body
         problem, AIAA 4th Aerospace Sciences Meeting, Los
         Angeles, California, June 27-29, 1966, AIAA Paper
         No. 66-411.

Chapter 3
OTHER SELECTED PROBLEMS

3.1. Introductory remarks.

T h e   s c o p e   o f   t h e   c h a p t e r: 1.
axisymmetric body whose generatrix has a discontinuous slope ;
2. Asymmetric plane body. We shall stay within the frame of
classical gasdynamics. The more, we shall base on the system of
equations (2.48), as derived in the paragraph 2.5.

These two problems do not cover all the
a p p l i c a t i o n s of the MOIR, as known so far.

In [1] already - application to subsonic flow
past ellipses and ellpsoids, as well as that to sonic flow past
ellipses - has been reported.

In [2] - application of the MOIR to : 1) subsonic flow past symmetric plane and axisymmetric bodies ; 2)
flow in Laval nozzle ; 3) conical flow with attached shock waves;
4) asymmetric flow around axisymmetric bodies (the case of <u>three</u>
independent variables) - are mentioned.

In [3] - some chapters are devoted to flows
with radiation, thermodynamically unstable flows, flows of
viscous fluids - to call but a few.

We were concerned-to mention the most "untypical"
application - e.g. with bodies with gaseous protective layer [4],

and with bodies possessing oscillating surface $\boxed{5}$ – both in
supersonic flow.

The two abovementioned problems were selected for
presentation in this course, because they might call the atten-
tion of the audience to some rather severe and fundamental ques-
tions arising sometimes when applying the MOIR.

3.2. Specific shapes of the bodies.

3.2.1. Computation of flow on axisymmetric bodies, which are smooth in their front parts, but whose generatrixes have not continuous slopes (Fig. 3.1.a).

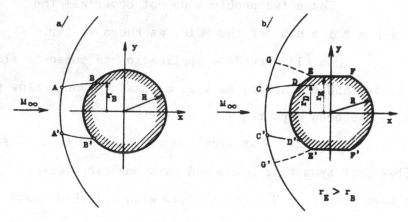

Fig. 3.1.

In connection with computation of flow around such
bodies, a problem arises, posed in $\boxed{6}$, concerning the location
of the sonic point on the body, wherein the regularity condition
has to be satisfied. There are two possibilities : the point is
located either at D (bodies with "sufficiently" long front

parts), or at the corner E.

The approach to computation of the flow would be different in each case. Location of the sonic point at D - calls for formulation of the mathematical problem exactly as it was described in the former chapter ; location at E - calls for another complementary condition :

$$ v_b \Big|_{S=S_E} = a_* \, , \qquad\qquad (3.1) $$

which must be accepted instead of (2.57).

Opinions which can be found in the literature are rather in favor of location at E (see references in [6]).

Possibilities of theoretical solution of this dilemma are rather vague. The most obvious methods of attack would be :

a) investigation of stability of two flows, corresponding to the both locations of the sonic point (not very promising : both may turn out to be stable with respect to small disturbations) ;

b) "stationarised" numerical approach - in sense of solving by finite differences the system of equations of classical gasdynamics (like in [11] or [12]), or even that of dynamics of viscous heat-conducting gas (like e.g. in [13]). Such an approach is however, hopeless - except when having the biggest computers at the disposal.

Experimental solution of the dilemma is presented in $\begin{bmatrix}6\end{bmatrix}$, wherefrom the Figs. 3.2. and 3.3. are taken. It is in favor of location of the sonic point at D.

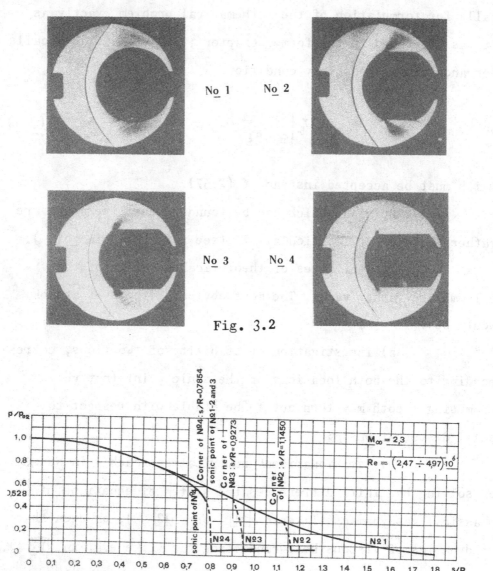

Fig. 3.2

Fig. 3.3

The conclusion from the computational viewpoint are obvious.

3.2.2. Body with conical front part [7] .

Axisymmetric bodies with conical front parts as well as plane bodies whose front parts form a wedge (Fig.3.4) can be relatively easy treated by means of the system of equations as developed in the paragraph 2.5.

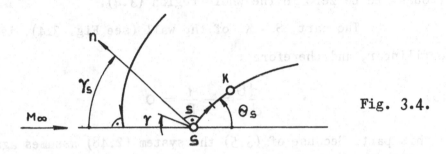

Fig. 3.4.

In this case Θ is accepted as the independent variable. Multiplying the system (2.48) by $\dfrac{ds}{d\Theta}$ and then substituting

$$\frac{ds}{d\Theta} = -R = 0 \tag{3.2}$$

in the terms, still containing this derivative after reduction of ds , we get the following system of equations :

$$\left.\begin{array}{c} A\dfrac{d\varepsilon}{d\Theta} + B\dfrac{d\nu_b}{d\Theta} + C\dfrac{d\delta}{d\Theta} + \delta \cdot D_2 = 0 \\[2mm] E\dfrac{d\varepsilon}{d\Theta} \qquad + F\dfrac{d\delta}{d\Theta} + \delta \cdot G_2 = 0 \\[2mm] \dfrac{d\delta}{d\Theta} + \delta \cdot L_1 = 0 \end{array}\right\} \tag{3.3}$$

This system is to be integrated with the usual initial conditions, as corresponding to the stagnation point, in the limits:

(3.4) $\dfrac{\pi}{2} \geqslant \theta \geqslant \theta_s$; $(0 \leqslant \gamma \leqslant \gamma_s)$.

The value of the velocity v_b at the upper limit θ_s of the integration represents somehow the measure of the inadequacy of the whole mathematical model of the flow, as it obvious‑ ly ought to be zero in the whole region (3.4).

The part S – K of the wall (see Fig. 3.4), is rectilinear, and therefore :

(3.5)   . $\dfrac{d\theta}{ds} = -\dfrac{1}{R} = 0$

on this part. Because of (3.5) the system (2.48) assumes again a special form :

(3.6) $\begin{cases} A\,\dfrac{d\varepsilon}{ds} + B\,\dfrac{dv_b}{ds} + C\,\dfrac{d\delta}{ds} = D_1 ; \\[2mm] E\,\dfrac{d\varepsilon}{ds} \qquad\quad + F\,\dfrac{d\delta}{ds} = G_1 ; \\[2mm] \qquad\qquad\qquad\quad \dfrac{d\delta}{ds} = L_1 . \end{cases}$

It is to be integrated from S to K , with initial conditions represented by final values as obtained from the integration of the previous system (3.3) at

(3.7) $\theta = \theta_s$.

If the wall is curvilinear "above" K, integration "switches" to the full system (2.48) with proper initial conditions, stemming again from integration of the system (3.6).

If the wall has an edge at K , this point is

Fig. 3.5.

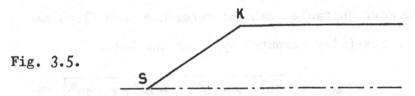

accepted as the critical point (Fig. 3.5).

If the rectilinear part of the wall (plane) or of the generatrix (axisymmetric) is absent, so that two curvilinear parts meet at the stagnation point, the system (3.6) is omitted, of course, and the integration switches from the system (3.3) in the upper limit (3.7) immediately to the full system (2.48).

Otherwise, the solution of the direct problem for such "special" shapes of the bodies does not involve any further difficulties.

A series of systematic computations has been performed [8] , and the body shapes investigated are reproduced in the Fig. 3.6. The series includes flat-nosed cylinders (case I ; $\gamma = 90°$).

A typical set of numerical results is shown in the Fig. 3.7.

3.2.3. Concave bodies.

Flow around such bodies obviously can not be computed by the use either of the system (2.48) or its variants, because of the accepted frame of reference n, s , which does not provide the unique mapping of the plane x, y into the plane n, s (many pairs n, s, may correspond to one point of the field). In the more suitable frame of reference such flows may be, however, successfully computed by using the MOIR.

	M_∞	$\gamma = 90°$	$\gamma = 80°$	$\gamma = 70°$	$\gamma = 60°$
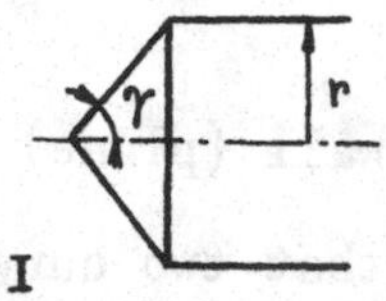 I	3	0.606	0.450	0.289	0.129
	5	0.510	0.354	0.198	
	10	0.469	0.314	0.160	
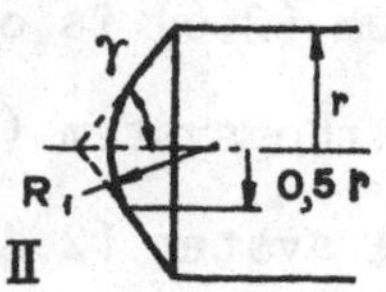 II	3		0.496	0.388	
	5		0.399		
	10		0.358	0.252	
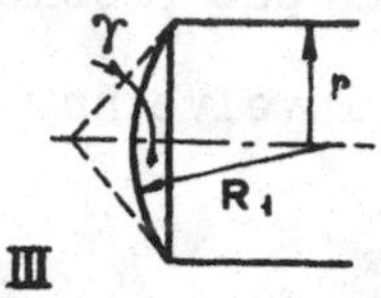 III	3		0.536	0.475	
	5		0.441	0.374	
	10		0.333		
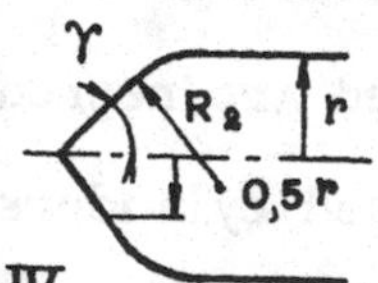 IV	3			0.189	
	5			0.122	
	10	0.316	0.201		
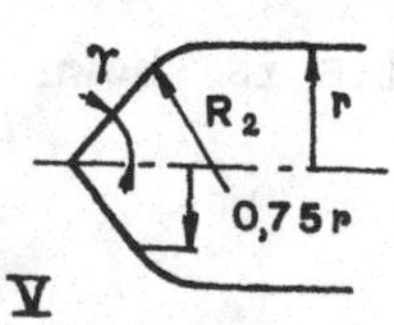 V	3				
	5				
	10		0.261	0.128	

Fig. 3.6.

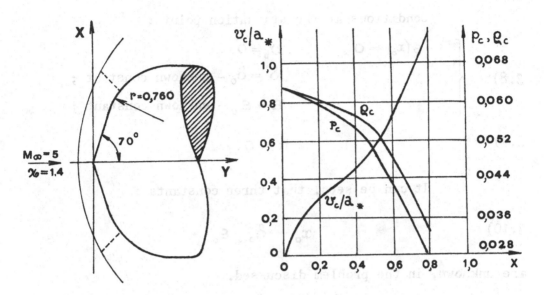

Fig. 3.7.

3.3. The asymmetric problem.

Note. In presentation of the asymmetric plane direct problem of supersonic flow around a blunt body (see Fig. 2.3) we will refer mainly to our paper [9], wherein such a problem was treated.

3.3.1. The system of equations.

As we deal with the direct problem, the system of equations is that derived in the paragraph 2.6.

3.3.2. Conditions.

Position of the stagnation point is unknown in advance in this case. Let's denote it by x_o.

The location of the both critical points is unknown, too, as it was discussed in the par. 2.6.

Conditions at the stagnation point :

(3.8) $\left\{\begin{array}{l} s(x_o) = 0 \qquad\qquad v_b = 0 \\[2mm] \qquad\qquad\qquad \delta = \delta_o \text{ - unknown constant ;} \\[2mm] \qquad\qquad\qquad \varepsilon = \varepsilon_o \text{ - unknown constant ;} \end{array}\right.$

(3.9) $$n = 0.$$

It can be seen, that three constants :

(3.10) $$x_o \, , \; \delta_o \, , \; \varepsilon_o$$

are unknown, in the problem discussed.

Two complementary conditions stem from the regularity requirements and are identical with (2.57) :

(3.11) $$D - CL - \frac{A}{E}(G - FL) \Big|_{v_b = \pm a} = 0$$

The third, lacking condition stems from the following assumption : the m a x i m u m e n t r o p y s t r e a m-
l i n e w e t s t h e b o d y (Fig. 3.8).

The mathematical formulation of the condition following from this assumption, can be accepted as

(3.12) $$\varepsilon_{int} = \frac{\pi}{2}$$

the symbol ε_{int} denoting the shock wave angle at the intersection point of the stagnation streamline and the shock wave.

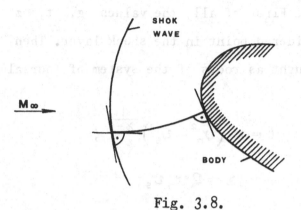

SHOK
WAVE

M_∞

BODY

Fig. 3.8.

The audience is referred to two papers, mentioned
in the previous Chapter, as far as discussion and critics of
this assumption is concerned. We do not wish to take sides as
far as the validity of this assumption is concerned – in the
course of these lectures : our only object is to emphasize the
need of <u>any</u> third assumption, caused by the appearence of <u>three</u>
unknown constants in the problem.

This defect (as it must be treated as a defect) is
common to a multitude of applications of the MOIR.

3.3.3. Computing of the stagnation streamline.

The equation of a streamline :

$$\frac{dn}{ds} = \frac{v_n}{v_s}\left(1 + \frac{n}{R}\right) \qquad (3.13)$$

must be included, in order to compute the stagnation streamline,
starting from the stagnation point (3.9).

The velocity components v_n, v_s are calculated
consistently with the accepted "scheme of approximation" – in

the following manner. First of all, the values g, t, z are
computed at the considered point in the shock layer. Then v_n,
v_s, ϱ , p, are sought as roots of the system of four algebraic
equations :

$$t = v_s \left(v_n^2 - v_s^2 \right)^{\frac{1}{\varkappa - 1}} ;$$

$$z = \varrho \, v_n \, v_s ;$$

$$g = \varrho \, v_n^2 + k \, p ;$$

$$\frac{p}{\varrho} = 1 - v_n^2 - v_s^2 .$$

This is done by a suitable trial and error procedure.

3.3.4. Iteration technique,

is based on three "error indicators" E_1, E_2, E_3, the third
one being defined as follows, accordingly to (3.12) :

(3.14) $$E_3 = \varepsilon_{int} - \frac{\pi}{2}$$

The two remaining ones are defined at the Fig. 3.9.

All three of them are evaluated in course of numerical integration of the system of four differential equations
(2.48), (3.13).

In order to evaluate E_1, the system (2.48) of
three differential equations is numerically integrated for a
set of arbitrarily chosen initial values x_o , δ_o , ε_o with
a positive step Δs , until the velocity modulus $|v_b|$ becomes

equal to a prescribed quantity, a little lower than a_* (Fig. 3.9 a, b), or until the numerator $\left[D - CL - \dfrac{A}{E}(G - FL)\right]$ becomes equal or less than zero (Fig. 3.9.c). The integration pauses then, and E_1 is evaluated by the use of linear interpolation, or extrapolation, as shown schematically on the Fig. 3.9.

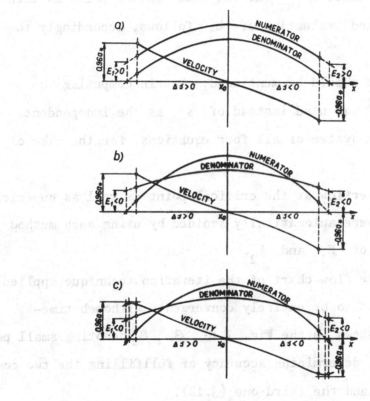

Fig. 3.9.

The same is true as far as evaluation of the second error indicator E_2 is concerned. Of course the integration is performed with negative step ($\Delta S < 0$) in this case.

In order to evaluate E_3 the system of four differential equations is integrated, i.e. (2.48), (3.13), starting
from the stagnation point, until the computed stagnation streamline reaches the wave :

(3.15)
$$n \geqslant \delta$$

The shock wave angle ε_{int} at the intersection point is then
interpolated, and evaluation of E_3 follows, accordingly to
(3.13).

It should be mentioned, that in computing the
streamline, n was used instead of s as the independent
variable in the system of all four equations, for the sake of
convenience.

Overflow at the critical point as well as numerical
instabilities were automatically avoided by using such method
of calculation of E_1 and E_2.

The flow chart of the iteration technique applied,
which turned out to be entirely convergent, although time-
consuming, is shown in the Fig. 3.10, β_1 , β_3 denoting small positive numbers, determining accuracy of fullfilling the two conditions (3.11) and the third one (3.12).

3.3.5. Results.

Some results computed for $M_\infty = 3$; $\varkappa = 1,4$ and
a prolate elliptical profile of axes ratio 4, taken from [9] ,
are represented in the picture 3.11.

The computations took ca 50^{hrs} on a GIER computer.

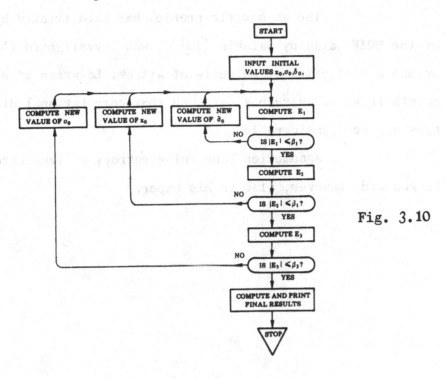

Fig. 3.10

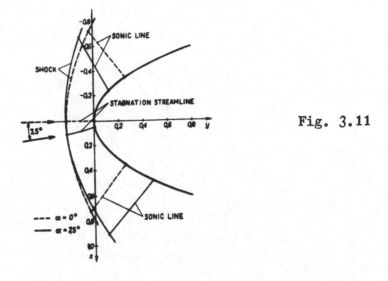

Fig. 3.11

3.3.6. Final remark.

The asymmetric problem has been treated by means of the MOIR also by Bazshin [10] , who investigated flow around a flat plate at an angle of attack. Location of sonic points is known in such a case, so that computational difficulties are not considerable.

Assumption concerning entropy is implicitely introduced, however, also in his paper.

References

[1]   A.A. Dorodnicyn, A contribution to the solution of mixed problems... see ref. 1.9.

[2]   О.М. Белоцерковский, П.П. Чушкин, Численный метод интегральных соотношений, Журнал мат., мат.физ., 2 (1962) , Ио.5, 731-759.

[3]   О.М. Белоперковский (ред.), Обтекание затупленных тел сверхзвуковым потоком газа, В.Ц. АН СССР – 1967.

[4]   W.J. Prosnak, J.M. Breiter, Computation of flow around a blunt-body with a gaseous protective layer. Fluid Dyn. Trans., 3 (1967), 565-585.

[5]   W.J. Prosnak, W. Fiszdon, A method for determination of hypersonic flow about a blunt body with oscillating surface. Fluid Dyn. Trans., 3 (1967), 587-605.

[6]   W.J. Prosnak, J. Stasiak, Experimental investigation of the sonic line on certain axisymmetric blunt bodies. IBTP – Reports 2/1970 (Polish Academy of Sci. – Warsaw.)

[7]   E. Luczywek, Analiza metody zwiazkow calkowych w zastosowaniu do badania oplywu cial tepych naddzwiekowym strumieniem gazu (PhD Thesis ; Politechnika Warszawska 1966).

[8]   E. Luczywek, Determination of supersonic flow around blunt bodies by the use of the MOIR (in Polish). IBTP – Reports, 10/1967 (Polish Acad. of Sci. – Warsaw).

[9]   W.J. Prosnak, E. Luczywek, The direct asymmetric hyper-
       sonic blunt–body problem, proceedings of the 4th
       ICAS Congress, Paris, 1964, 911– 930.

[10]   А.Г. Бажин, К расчету обтекания сверхзвуковым
       потоком газа плоской пластинки с
       неприсоелиненным скачком уплотнения. Инж.
       журн., III (1963), 2, 222–227.

[11]   C.P. Kentzer, Transonic flows past a circular cylinder
       (Purdue University, Lafayette, Indiana, USA).
       Preprint Presented at the Section on Numerical
       Methods in Gasdynamics of the Second International
       Colloquium on Gasdynamics of Explosion and Reacting
       Systems, Novosibirsk, Aug. 19–23, 1969.

[12]   C.P. Kentzer, Computation of time dependent flows on an
       infinite domain. AIAA Paper No. 70–45. Presented
       at the AIAA Sth Aerospace Sciences Meeting, New
       York, Jan. 19–21, 1970.

[13]   S.M. Scala, and P. Gordon, Solution of the time – depen-
       dent Navier – Stokes equations for the flow around
       a circular cylinder. AIAA Journ., Vol. 6., No. 5,
       pp. 815–822 (1968).

Table of Contents

Printed in the United States
By Bookmasters